REPORT

Upgrading the Extender

Which Options Are Cost-Effective for Modernizing the KC-10?

Anthony D. Rosello • Sean Bednarz • David T. Orletsky • Michael Kennedy
Fred Timson • Chuck Stelzner • Katherine M. Calef

Prepared for the United States Air Force

Approved for public release; distribution unlimited

RAND PROJECT AIR FORCE

The research described in this report was sponsored by the United States Air Force under Contract FA7014-06-C-0001. Further information may be obtained from the Strategic Planning Division, Directorate of Plans, Hq USAF.

Library of Congress Cataloging-in-Publication Data is available for this publication.

ISBN 978-0-8330-5109-7

The RAND Corporation is a nonprofit institution that helps improve policy and decisionmaking through research and analysis. RAND's publications do not necessarily reflect the opinions of its research clients and sponsors.

RAND® is a registered trademark.

Cover: U.S. Air Force photo by MSgt Robert Wieland

Published 2011 by the RAND Corporation
1776 Main Street, P.O. Box 2138, Santa Monica, CA 90407-2138
1200 South Hayes Street, Arlington, VA 22202-5050
4570 Fifth Avenue, Suite 600, Pittsburgh, PA 15213-2665
RAND URL: http://www.rand.org/
To order RAND documents or to obtain additional information, contact
Distribution Services: Telephone: (310) 451-7002;
Fax: (310) 451-6915; Email: order@rand.org

Preface

The KC-10 "Extender" tanker aircraft has been in operation with the U.S. Air Force since 1981 without significant modernization. At the request of the Office of the Assistant Secretary of the Air Force for Acquisition, Global Reach Programs, the RAND Corporation conducted a cost-effectiveness analysis of modernizing the KC-10 in the areas of avionics (communication, navigation, and surveillance [CNS] capabilities for air traffic management [ATM]), night-vision imaging system (NVIS) compatibility, command and control (C2, specifically, data-link capability), additional multipoint refueling capability, defensive protection, and reliability and safety upgrades.

The avionics (CNS/ATM) analysis is documented in *Assessing the Cost-Effectiveness of Modernizing the KC-10 to Meet Global Air Traffic Management Mandates* (Rosello et al., 2009). This work presents the analysis of the remaining areas in the context of quantitative benefits to KC-10 wartime missions and describes other salient considerations for modifying the KC-10.

This research was sponsored by Maj Gen Randal D. Fullhart, director, Global Reach Programs, Office of the Assistant Secretary of the Air Force for Acquisition, Headquarters U.S. Air Force, and conducted within the Force Modernization and Employment Program of RAND Project AIR FORCE for a fiscal year (FY) 2008 project titled "KC-10 Modernization Roadmap."

RAND Project AIR FORCE

RAND Project AIR FORCE (PAF), a division of the RAND Corporation, is the U.S. Air Force's federally funded research and development center for studies and analyses. PAF provides the Air Force with independent analyses of policy alternatives affecting the development, employment, combat readiness, and support of current and future aerospace forces. Research is conducted in four programs: Force Modernization and Employment; Manpower, Personnel, and Training; Resource Management; and Strategy and Doctrine.

Additional information about PAF is available on our website:
http://www.rand.org/paf/

Contents

Figures

Tables

Summary

The Air Force asked RAND Project AIR FORCE to undertake a study to provide objective insight into the cost-effectiveness of modernizing the KC-10 "Extender" air refueling (AR) tanker aircraft. The study analyzed the cost-effectiveness of modernizing the KC-10 in the areas of avionics (CNS capabilities for ATM), NVIS compatibility, C2 (specifically, tactical data-link [TDL] capability), additional multipoint refueling capability, defensive protection, and reliability and safety upgrades.

We assessed the cost-effectiveness of the various modernization options by estimating each option's total life-cycle cost and comparing that cost with its quantitative benefit. The quantitative benefit of each option was determined by valuing the number of tanker aircraft saved because of the KC-10's increased wartime mission effectiveness and the effects on peacetime operating costs after modernization. In some cases, modernization options provide benefits that do not directly affect the cost or effectiveness of the KC-10 but, rather, improve commanders' operational flexibility in employing the KC-10 or improve the effectiveness of other weapon systems. In such cases, we highlight these additional benefits or implications. (See pp. 5–7.)

The context for evaluating changes to peacetime operating costs is 11 years of KC-10 operational flying data. To evaluate the impact on executing wartime missions, we used representative missions vetted in RAND's KC-135 recapitalization analysis of alternatives (AoA) (see Kennedy et al., 2006) and the *Mobility Capabilities Study* (DoD and JCS, 2005), supported by tanker doctrine. These missions include homeland defense, air bridge, national reserve, global strike, theater employment, deployment, and Operations Plan (OPLAN) 8010 (Strategic Deterrence and Global Strike). We use the mission title *air bridge* to capture the missions of global strike: air bridge, OPLAN 8010, and national reserve. Although these three missions vary in their overall military purpose and goals, they are very similar from the perspective of the tanker operations required to support them. Thus, our analysis modeled requirements and matched the selected modernization options to four broad mission types: homeland defense, theater employment, deployment, and air bridge. (See pp. 9–15.)

The modernization options provide benefits to operations in different ways for different missions. Not all options benefit all missions. For example, defensive systems may allow tankers to base closer to AR locations and to conduct AR closer to threats than without the systems. However, defensive systems do nothing to improve the rate at which receivers cycle across the boom or baskets. The benefits provided by each modernization option through various types of missions are shown in Table S.1. (See pp. 15–18.)

Using a value for the KC-10 based on cost research in RAND's KC-135 recapitalization AoA (see Kennedy et al., 2006), we determined the value of improvements in effectiveness,

Table S.1
Modernization Options, Missions, and Tanker Efficiency Benefits

Modernization Option	Benefit and Mission			
	AR Orbits Farther Forward	Tanker Bases Closer to AR Orbits	More Efficient Planning and Operations	Faster Receiver Cycle Times
TDL	Employment		Homeland defense Employment Deployment Air bridge	
Additional multipoint refueling capability				Employment Deployment
Defensive systems	Employment	Employment		
NVIS-compatible lighting				Employment

reliability, and safety, which are shown in Table S.2. We evaluated changes in effectiveness for each of the system modernization options. Changes in reliability and safety were not explicitly analyzed for each modernization option but can be used to determine the price that the Air Force should be willing to pay for these improvements. (See pp. 15–17.)

After examining the costs and benefits of each of the modernization options,[1] we compared their relative merits, ordering the options by cost-effectiveness ratio and the ratio of improvement in wartime effectiveness to the modernization cost of each option, including any change to operating costs. The cost-effectiveness ratio shows not only how the options compare in terms of best value per dollar but also at what point the returns on modernization spending begin to decrease. This approach of comparing the options does not capture costs or benefits that are inherently not quantifiable but may be important considerations when deciding to upgrade the KC-10 fleet. In those cases, we review the important considerations for each of the options. (See pp. 19–28.)

The modernization options in order of the greatest to least cost-effectiveness are adding a TDL, CNS/ATM, additional multipoint refueling, defensive systems, and NVIS-compatible lighting. The first three—TDL, CNS/ATM, and additional multipoint refueling—all have

Table S.2
Value of Changes in Effectiveness, Reliability, and Safety

Change	Value (FY 2009 $ millions)
1% effectiveness increase	2.9
1% not-mission-capable rate decrease	2.5
1% depot-possessed rate decrease	2.4
0.1% attrition rate reduction	2.8

---

[1]   We estimated the costs of each modernization option independently. If some options are implemented simultaneously, there could be reduced costs because of overlapping access requirements (i.e, TDL and CNS/ATM both require access to the cockpit). However, given historical cost growth of programs and uncertainty in cost estimates, our approach is conservative.

positive net present values (NPVs), meaning that the overall benefit is greater than the cost to procure these upgrades. Upgrades for defensive systems could be cost-effective (i.e., have a positive NPV) if either (1) KC-10s are used heavily for employment missions and can be based significantly closer to AR orbit locations or (2) KC-10s are used to offset C-17s in an airlift role. NVIS-compatible lighting is not cost-effective for the KC-10. Table S.3 shows the cost and benefit of each of the modernization options. The benefits in Table S.3 are based on the average of two mission mixes that represent different ways in which the KC-10 could be used in wartime: one weighted toward theater employment missions, the other weighted toward deploying fighter-sized aircraft to theater. (See pp. 73–75.)

Of the options, adding a TDL to the KC-10 has the greatest cost-effectiveness ratio. The data link is a relatively inexpensive upgrade compared with the other options. Among other capabilities, a TDL would provide the KC-10 with position and mission information on receiver aircraft without relying on voice communication. This information would allow the reduction of planned overlap times and facilitate faster rendezvous with receiver aircraft, making the KC-10 more effective. (See pp. 29–37.)

Modifying the KC-10 avionics upgrades to be compliant with upcoming worldwide equipage mandates has the next-highest cost-effectiveness ratio. Most of the CNS/ATM upgrade benefit is the avoidance of fuel penalties because the equipment is mandated to access the most fuel-efficient altitudes. Under a broad range of assumptions regarding savings and fuel costs, the CNS/ATM upgrade is cost-effective based on peacetime savings only. However, the findings show that, even under a worst-case cost scenario, the savings resulting from KC-10 fleet modernization would exceed the cost of the upgrade long before the fleet is retired in 2045. That research is detailed in *Assessing the Cost-Effectiveness of Modernizing the KC-10 to Meet Global Air Traffic Management Mandates* (Rosello et al., 2009).

Additional multipoint refueling capability increases effectiveness primarily in the employment mission when refueling multiple strike and air defense aircraft, with a lesser benefit for the deployment mission. The number of aircraft allowed to fly in refueling formation with the tanker limits the potential effectiveness gain from multipoint in the deployment mission. Our research suggests that eight receivers is a reasonable maximum, and that number is the basis

Table S.3
Costs and Average Benefits of Each Modernization Option

Modernization Option	FY 2009 $ Millions/ Total Aircraft Inventory	
	Cost	Benefit
TDL	0.7	6.5
Additional multipoint refueling capability	4.2	11.6
Defensive systems	21.4	10.2
NVIS-compatible lighting	3.6	0.1
CNS/ATM	7.5	26.1

NOTE: All costs and benefits are presented in terms of millions of FY 2009 dollars per aircraft. We express this as FY 2009 $ millions/total aircraft inventory to indicate that these per-aircraft values were calculated using the entire KC-10 fleet size.

for the benefit presented here. Values for six and 12 receivers are also presented in Chapter Six. (See pp. 39–50.)

Defensive system upgrades are cost-effective only if these systems allow the KC-10 to be based significantly closer to wartime operational AR locations than established in planning documents and practiced in recent conflicts. Defensive systems may also be cost-effective by allowing the KC-10 to be used more in an airlift role, thus freeing a number of large defensive system–equipped airlifters (C-17s or C-5s, for example) to conduct other missions for which they are best suited. Our values are based on adding the proposed defensive system suite and basing the KC-10s 200 nautical miles (nm) closer to AR orbits. The rationale for the 200 nm stems from basing locations in Operation Iraqi Freedom. At these values, the cost of the upgrade would be greater than the value of its benefit. In the case of defensive systems and closer basing, tanker experts and decisionmakers can trade off system cost, the extent of the upgrade, and how close they are willing to base the aircraft. The parametric analysis in Chapter Seven can help determine the trade-offs for other costs and distances. (See pp. 51–66.)

Retrofitting the KC-10 with NVIS-compatible lighting is not cost-effective because it does little to make the tanker more effective. Air Force testing and empirical safety data suggest minimal improvement in tanker mission effectiveness with NVIS-compatible lighting. (See pp. 67–71.)

Figure S.1 shows each of the modernization options in order of their cost-effectiveness in a cumulative plot of costs and benefits. As the figure shows, TDL, CNS/ATM, and multipoint refueling capability each provide more benefit than cost, and defensive systems and NVIS-compatible lighting cost more than the benefit they provide. As a package, if all the upgrades were pursued, the overall benefit would be greater than the overall cost of all the upgrades. (See pp. 73–75.)

Figure S.1
Cumulative Cost-Benefit Curve of Modernization Options

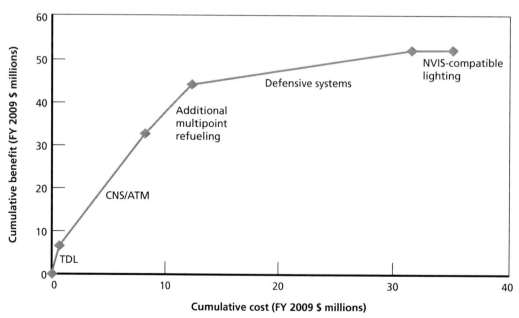

RAND *TR901-S.1*

Acknowledgments

We are grateful to many individuals in the U.S. Air Force and the U.S. government who contributed to this study. Maj Gen Randal D. Fullhart, director, Global Reach Programs, Office of the Assistant Secretary of the Air Force for Acquisition, sponsored this work; his predecessor in that office, Maj Gen David S. "Scott" Gray, was the sponsor until November 2008. They provided support and guidance throughout the study. Our action officers in the Global Reach Programs Mobility Division, Lt Col Brian Jonasen, Lt Col Eugene Croft, Allan Haenisch, and Tod Beatrice, consistently provided helpful input and guidance. Lt Col Matt Bonavita at Air Force Materiel Command was instrumental in initiating this effort.

At Air Mobility Command (AMC), we benefited from the input, critical review, support, and coordination provided by our day-to-day action officer, John O'Neill, AMC/A9 (Analysis, Assessments, and Lessons Learned Directorate). We also thank Dave Merrill and Jim Donovan at AMC/A9 for their support, guidance, and feedback. We are grateful to the many others at AMC who provided data, information, and critical reviews of our work. Lt Col James Craft, David Ziegler, and James Rummer at AMC/A5Q (Plans, Programs, and Requirements Directorate) shared their interest, corporate knowledge, and suggestions, all of which improved this analysis. We also appreciate the input and participation of Maj Thomas Kanak and Lt Col James Pena at AMC/A3 (Air, Space, and Information Operations Directorate).

Lt Col Dave "Fuze" Jeter facilitated our visit and interaction with the Royal Netherlands Air Force, where Col J. W. M. Scheepers and his team provided insight into the Dutch upgrades of the KDC-10 fleet. We also thank Paul Morris and the United Kingdom's Royal Air Force team for hosting us and sharing their tanker and data-link experience. Lt Col Vernon Conaway shared his theater safety experience, and Zachary Cooper and Randy Smejkal from the KC-10 Program Office at Tinker Air Force Base (AFB), Oklahoma, shared their input and corporate knowledge of the KC-10. Maj Steve Haren and Maj Ryan Vander Veen also hosted an exceptionally professional and informative visit to McGuire AFB, New Jersey.

This study also benefited from the contributions of Peter White from FedEx, who hosted an informative visit on his organization's MD-10 program; Gary Liberson at Lockheed Martin, who organized and hosted a day of discussion on future airpower operations; and Jeremiah Gertler of the Congressional Research Service, who thoughtfully reviewed our analysis.

Finally, we thank of our colleagues at the RAND Corporation: Carl Rhodes, John Tonkinson, and Ken Munson, for their careful and insightful reviews of an earlier version of this document; Andrew Hoehn, RAND vice president and director of PAF, for his continued support; Donald Stevens, our program director, for his guidance and input; Jane Siegel, for preparing and formatting an earlier version of this technical report; and Lauren Skrabala, for her careful editing.

Abbreviations

AFB	Air Force base
AMC	Air Mobility Command
AMC/A3	Air Mobility Command Air, Space, and Information Operations Directorate
AMC/A5Q	Air Mobility Command Plans, Programs, and Requirements Directorate
AMC/A9	Air Mobility Command Analysis, Assessments, and Lessons Learned Directorate
AoA	analysis of alternatives
AR	air refueling
ARO	air refueling operator
ATM	air traffic management
AUPC	average unit procurement cost
AWACS	Airborne Warning and Control System
BLOS	beyond line of sight
C2	command and control
C4	command, control, communication, and computers
CAF	combat air force
CAP	combat air patrol
CAS	close air support
CNS	communication, navigation, and surveillance
COCOM	combatant command
CONOPS	concept of operations
DHS	U.S. Department of Homeland Security

FY	fiscal year
ICD	initial capabilities document
IR	infrared
ISR	intelligence, surveillance, and reconnaissance
ISO	International Organization for Standardization
JRE	Joint Range Extension
JTIDS	Joint Tactical Information Distribution System
LAIRCM	Large Aircraft Infrared Countermeasures System
LOS	line of sight
MANPADS	man-portable air defense system
MIDS	Multifunctional Information Distribution System
nm	nautical mile
NMC	not mission capable
NPV	net present value
NVG	night-vision goggles
NVIS	night-vision imaging system
OIF	Operation Iraqi Freedom
OMB	Office of Management and Budget
OPLAN	operations plan
PAF	RAND Project AIR FORCE
PB	President's Budget
RF	radio frequency
SADL	situation awareness data link
SAM	surface-to-air missile
SIPRNet	Secret Internet Protocol Router Network
STU	secure telephone unit
TADIL-J	Tactical Digital Information Link J
TAI	total aircraft inventory
TDL	tactical data link
UCAV	unmanned combat aerial vehicle

UHF	ultrahigh frequency
USAFRICOM	U.S. Africa Command
USCENTCOM	U.S. Central Command
USEUCOM	U.S. European Command
USNORTHCOM	U.S. Northern Command
USPACOM	U.S. Pacific Command
USSOUTHCOM	U.S. Southern Command
WUC	work unit code

CHAPTER ONE

Introduction

The U.S. Air Force asked RAND Project AIR FORCE to provide objective insight into the cost-effectiveness of modernizing the KC-10 "Extender" air refueling (AR) tanker aircraft. The study evaluated KC-10 modernization options in the areas of areas of avionics (communication, navigation, and surveillance [CNS] capabilities for air traffic management [ATM]), night-vision imaging system (NVIS) compatibility, command and control (C2, specifically, data-link capability), additional multipoint refueling capability, defensive protection, and reliability and safety upgrades.

General Approach

In the study, we assessed the cost-effectiveness of various modernization options by estimating each option's total life-cycle cost and comparing that cost with its quantitative benefit. We determined the quantitative benefit of each option by valuing the number of tanker aircraft saved because of the KC-10's increased effectiveness after modernization. In some cases, modernization options provide benefits that do not directly affect the cost or effectiveness of the KC-10 but, rather, improve commanders' operational flexibility in employing the KC-10 or the effectiveness of other weapon systems. In these cases, we highlight the additional benefits and implications.

Establishing Modernization Options

Our first task was to define the set of modernization options that we would investigate for the KC-10. We referenced the Air Force's *Initial Capabilities Document for Aerial Refueling,* published in 2005, which highlighted seven areas that the Air Force viewed as deficient in AR capability that could be addressed though aircraft modernization. The seven areas identified in the ICD were (1) NVIS compatibility; (2) CNS/ATM; (3) multipoint refueling; (4) boom/probe and drogue refueling; (5) command, control, communication, and computers (C4); (6) defensive protection; and (7) receiver capability. Not all of these deficiencies are applicable to the KC-10; the document referred broadly to the Air Force air refueling fleet, which includes both the KC-10 and KC-135 aircraft. Specifically, boom/probe and drogue refueling and receiver capability are not applicable to the KC-10. Next, we briefly describe the previously mentioned areas of deficiency and their applicability to the KC-10.

NVIS-compatible lighting refers to the installation of airplane lighting that is compatible with night-vision devices. With compatible lighting, the crew of the KC-10 and receiver aircraft can conduct night operations while wearing night-vision goggles (NVG) without airplane lighting interfering with their vision.

CNS/ATM refers to the communication, navigation, and surveillance avionics equipment required for ATM. CNS/ATM modernization involves an avionics upgrade to comply with upcoming global equipage mandates to facilitate more efficient civil air traffic control.

Multipoint refueling describes the capability of a tanker aircraft to simultaneously refuel two or more receiver aircraft. The KC-10 fleet currently includes a limited number of aircraft modified to perform multipoint refueling with additional fuel system plumbing and wing pods that contain hoses and drogues.

Boom/probe and drogue refueling describes the capability to conduct both current AR methods on a single flight. All KC-10s have this capability through the use of the boom assembly and centerline hose and drogue unit.

In the ICD (USAF, 2005c), *C4 for AR* refers broadly to improved connectivity with other aircraft as well as Air Force and joint information networks. In addition to standard aircraft radios, the KC-10 has limited connectivity through a satellite Iridium phone that is connected to the intercom system.

Defensive protection describes modernization that improves the survivability of tanker aircraft. Currently, the KC-10 does not have any systems to improve survivability against enemy threats.

Receiver capability refers to the aircraft's ability to receive fuel from another aircraft while airborne. All KC-10s are capable of receiving fuel while airborne.

The ICD broadly addressed AR capability without respect to the individual tanker platform. Some of the areas do not apply to the KC-10 because it already has the desired capabilities—specifically, boom/probe and drogue refueling and receiver capability. Therefore, we do not address improvements to the KC-10 in these two areas in this report.

The other areas *do* apply to the KC-10 and are addressed in our analysis. The assessment of CNS/ATM was a priority for the Air Force; the cost-effectiveness analysis of CNS/ATM is presented in *Assessing the Cost-Effectiveness of Modernizing the KC-10 to Meet Global Air Traffic Management Mandates* (Rosello et al., 2009). That analysis shows that CNS/ATM modernization to meet scheduled global equipage mandates is cost-effective based on peacetime fuel and flying-hour savings alone. Additionally, benefits to KC-10 wartime mission effectiveness and increased access to both airports and refueling training airspace favor the decision to modernize the KC-10 in this area. We estimated the cost to modernize the KC-10 fleet's avionics for compliance as between $400 million and $450 million and the cost avoidance over the life of the fleet from peacetime operations alone to exceed $1 billion (all estimates are fiscal year [FY] 2009 dollars).

In addition to the areas listed in the ICD, in this study, we also examined the reliability, safety, and maintainability of the KC-10. For each of these areas, we valued incremental improvements to inform spending decisions.

Determining the Benefits Provided

After establishing which modernization areas to examine, we determined how these options would improve the capability of the KC-10. Characterizing how modernization options benefit U.S. Air Force operations in general provides a consistent framework for quantifying and valuing these benefits.

The benefits generally fall into two categories: those that make the tanker more effective and those that provide a completely new capability for the tanker or supported aircraft. In cases in which options make the tanker more effective, we compare the number of tankers required to conduct missions both with and without modernization. We then relate the number of tankers saved to the estimated cost of the option, allowing a comparison of different options. On the other hand, for options that add new operational capability, we describe the capability and how it may be employed, along with a quantitative estimate of the level of tanker effort required for a given mission.

Expected Missions of the KC-10

To provide context for quantifying potential tanker savings and describing new capabilities resulting from modernization, we modeled tanker operations for a representative set of missions.

Missions assessed in this analysis are based on information in RAND's KC-135 recapitalization analysis of alternatives (AoA; see Kennedy et al., 2006), the *Mobility Capabilities Study* (DoD and JCS, 2005),[1] and tanker doctrine as found in Joint Publication 3-17 (JCS, 2009) and Air Force Doctrine Document 2-6 (2006).[2] The specific missions assessed were homeland defense, Operations Plan (OPLAN) 8010 (Strategic Deterrence and Global Strike), employment, deployment, air bridge, national reserve, and global strike. Each is discussed in more detail later in this report. We compare the number of tankers required to conduct the mission with and without the modernization options to determine each option's cost-effectiveness.

Profile of the KC-10 Fleet

The Air Force owns and operates a fleet of 59 KC-10 aircraft based at McGuire Air Force Base (AFB), New Jersey, and Travis AFB, California. The KC-10 has been in operation since 1981 and has not had a major upgrade in its lifetime.

In this report, we assess the costs and benefits of modernization based on Air Force operating the KC-10 fleet for the rest of its service life, assumed here to last until 2045. To determine changes to peacetime operating costs, we assume that KC-10s carry out missions every year equivalent to the average flying program between 1996 and 2006, in which the aircraft flew 950 hours per aircraft on a total aircraft inventory (TAI) basis at an average fuel burn of 18,900 lb (2,800 gallons) per hour.

Organization of This Report

Chapter Two describes in more detail the potential operational benefits that could be expected from KC-10 modernization. It addresses these benefits without regard to specific modernization options or missions and provides a framework for thinking about how tanker operations can be more effective or new capabilities fielded. Chapter Three presents the representative tanker missions used to evaluate the modernization options. Chapter Four outlines our methodologies for valuing benefits, estimating the cost of each of the options, and determining the valuation of improvements to KC-10 reliability, maintainability, and safety. Chapters Five

---

[1] The most recent mobility study, *Mobility Capabilities and Requirements Study 2016* (DoD and JCS, 2010), was released after we completed our analysis. The overall results of our study remain relevant, however, because the missions and mixes presented in that document were consistent with the previous study, tanker doctrine, and this report.

[2] Both doctrine documents provide excellent background on AR operations. For even more detail on AR, we recommend NATO's ATP-56B (2010).

through Eight detail the treatment and analysis of tactical data links (TDLs), multipoint refueling capability, defensive systems, and NVIS-compatible lighting upgrades, respectively.

CHAPTER TWO

Benefits of Modernization

Each KC-10 modernization option provides unique benefits for the military. Some modernization options would make the tankers more effective—that is, fewer tankers would be needed to conduct the same mission after they are modernized. Other modernization options provide benefits that do not reduce the number of tanker aircraft required but may bring a new capability to the Air Force or make other platforms more effective. This chapter describes the expected benefits from tanker modernization according to these two categories.

Benefits That Affect the Number of Tankers Required for Wartime Missions

Benefits detailed in this section are significant because they reduce the number of tankers required to execute wartime missions after modernization. Later in this report, we relate the modernization options to the benefits they provide for specific missions.

Aerial Refueling Orbits Farther Forward

Modernization may allow KC-10s to conduct AR farther forward than they do in current operations or in defense planning scenarios. By *farther forward*, we mean that the location where AR occurs is closer to the mission area of the receiver aircraft. It is important to note that moving the tanker orbit farther from the tanker base *increases* the tanker effort required. The benefit of moving the orbits forward comes from making tactical receiver aircraft more effective; the receiver aircraft will spend less time and fuel traveling between mission and refueling areas. The increase in the required number of tankers is a trade-off that provides the tactical aircraft with increased range or more time on station for orbiting missions.

Tanker Bases Closer to Aerial Refueling Orbits

Some modernization options contribute to allowing the KC-10 to base in locations previously not accessible because of threat conditions at airfields. If the KC-10 were able to base closer for operations, it would spend less time and fuel transiting to and from refueling locations; thus, fewer tanker aircraft would be required for a given operation. The reduction in the number of tankers needed to supply a given fuel demand results from three effects: less transit time to and from the AR location, which improves tanker cycle time; the ability to stay on station longer, which reduces the number of tankers required to support a continuous presence; and each tanker's capability to offload a greater amount of fuel once it arrives at the AR location.

It should be noted that tanker basing is not just a function of threat and threat-mitigation options. Rather, several other factors also contribute to aircraft and tanker basing decisions.

These factors include but are not limited to fuel availability, ramp space to park the aircraft, the location's ability to support the personnel required to operate the aircraft, and the relative ranges of the different aircraft in the theater. Nonetheless, the addition of defensive systems would provide the KC-10 with the protection that it currently lacks and equip it with the systems that are required on other U.S. Air Force aircraft for access to specific higher-threat locations.

More Efficient Planning and Operations

This benefit is twofold. First, it allows fewer tankers to conduct wartime missions by reducing the tankers' required lead time currently figured into most missions. Allowing tankers to safely "cut it closer" (by providing them with precise, real-time information about the location of receiver aircraft) will conserve tanker resources. Second, it increases tanker effectiveness by expediting rendezvous with receiver aircraft. By reducing these times, tanker aircraft will be able to shorten mission cycle time as well as reduce the amount of fuel they consume. In the case of expedited rendezvous, both tanker and receiver aircraft benefit from reduced cycle times and reduced fuel consumption.

Faster Receiver Cycle Times

Modernization options that allow receivers to complete refueling operations faster will reduce fuel consumption and flight time for both tankers and receivers. Additionally, in the case of multiple receiver refueling operations (for example, flights of fighters), faster receiver cycling will conserve the largest amount of fuel for the most fuel-critical receiver, leaving the flight as a whole with more fuel to conduct its assigned mission. The additional receiver fuel translates into more time on station for such missions as defensive counterair, as well as an increased radius for strike missions.

Other Military Mission Improvements

Other benefits of tanker modernization do not necessarily reduce the number of tankers required to perform a given mission but do yield benefits to the warfighter. These benefits sometimes make other weapon systems more effective, bring additional capability to the combined fleet, or provide additional operational tanker flexibility.

Reduced Attrition

Improved aircraft protection and situational awareness of other aircraft and threats may allow the KC-10 to operate with a reduced risk of aircraft loss or in higher-threat areas. This reduction in attrition stems from both improved safety in terms of accidental loss and mitigating losses from potential enemy threats.

Increased Mission Range or Time on Station for Receiver Aircraft

An additional tanker capability that moves AR locations farther forward would provide receiver aircraft with additional range to conduct missions (e.g., airstrikes or intelligence, surveillance, and reconnaissance [ISR]) or permit missions that require more time on orbit.

Airlift Augmentation

The KC-10 currently has a significant cargo-carrying capability through the use of its reinforced floor and side cargo door. It cannot carry large, outsized cargo because such items must be lifted to the door by a loader and then fit through the door. However, the KC-10 can carry up to 27 pallets of bulk cargo as long as the pallets are shaped properly. Despite the significant physical ability of the KC-10 to conduct airlift, airlift is a secondary mission, and the aircraft is not widely used in the cargo-carrying role. Modernization options that would allow greater use of the KC-10 as an airlifter could offset the requirement for some number of purpose-built airlifters in wartime planning (i.e., C-5s, C-17s, and C-130s). Another benefit of increased airlift capability is the flexibility offered to the Air Force of having a true dual-role aircraft.

Relay Augmentation

An increased data communication capability afforded to the tanker crew can also benefit other aircraft communication within line of sight (LOS) of the KC-10 by extending its range, increasing robustness, and translating different message protocols. For example, low-flying A-10 attack aircraft conducting close air support (CAS) missions often have limited LOS capability and share data with ground forces using the situation awareness data link (SADL). A tanker operating overhead, conducting AR, and equipped with data-link equipment could act as a relay between the A-10 other aircraft not visible to the A-10—and potentially back to a C2 center.

Summary

This chapter presented the potential benefits of the KC-10 modernization options in two categories: those that change the number of tankers required for wartime missions and those that improve other military missions. Benefits that change the number of tankers include moving AR orbits farther forward, allowing tankers to base closer to AR orbits, more efficient planning and operations, and faster receiver cycle times. Benefits that improve other military missions include reduced attrition, increased combat air forces (CAF) range or combat air patrol (CAP) coverage, airlift augmentation, and relay augmentation.

The next chapter presents the wartime missions of the KC-10 and relates those missions to the aforementioned upgrade options and benefits.

CHAPTER THREE

KC-10 Warfighting Missions

To evaluate the change in wartime effectiveness under different modernization options, we modeled KC-10 wartime mission operations. The goal of the modeling was to quantify the change in the number of tankers required to conduct this set of missions should the KC-10 be modernized. If a modernization option allows fewer tankers to conduct the same mission, the Air Force can determine how much it should be willing to pay for the upgrade based on the number of tankers saved. Missions assessed in this analysis are based on information in RAND's KC-135 recapitalization AoA (see Kennedy et al., 2006), the *Mobility Capabilities Study* (DoD and JCS, 2005), and tanker doctrine in Joint Publication 3-17 (JCS, 2009). The specific missions assessed were homeland defense, OPLAN 8010 (Strategic Deterrence and Global Strike), employment, deployment, air bridge, national reserve, and global strike.

Homeland Defense

The homeland defense tanker mission provides AR support in national airspace to fighter aircraft conducting defensive CAP and to Airborne Warning and Control System (AWACS) aircraft conducting radar surveillance to detect hostile aircraft and directing fighter aircraft responding to threats. Figure 3.1 illustrates the support provided to the fighter aircraft conducting the defensive CAP missions. The tankers are based 500 nautical miles (nm) from the CAP location. Once on station, they support a two-ship of F-15s until reaching bingo fuel and then returning to their departure location. Tankers must land at the departure base with appropriate reserve fuel, which we model as 10 percent of total fuel. We modeled AR support to AWACS, shown in Figure 3.2, where the tanker is based 600 nm from the AR location, where it will rendezvous with the AWACS and offload 95,000 lb of fuel before returning to base. Ninety percent of the homeland defense mission supports the fighters, and 10 percent supports the AWACS. The individual locations of the CAPs are assumed to be dispersed geographically, so one tanker is continuously required for each location.

Theater Employment

The theater employment mission is the most demanding mission for the KC-10 and consists of supporting a major combat operation from a deployed theater location and four AR locations. Six sub-missions make up the theater employment mission. These missions are support to air superiority, airborne electronic attack, close controlled strike, intratheater strike, long-range

Figure 3.1
Homeland Defense Tanker Support to Fighter CAPs

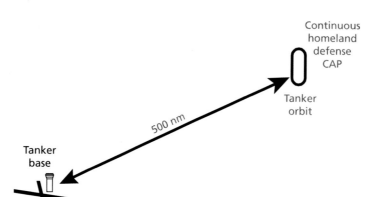

RAND *TR901-3.1*

Figure 3.2
Homeland Defense Tanker Support to AWACS

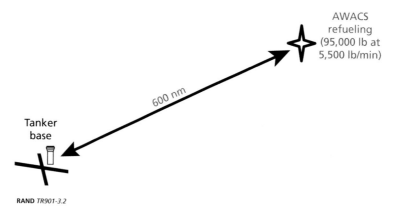

RAND *TR901-3.2*

strike, and large ISR platforms. The missions and profiles are based heavily on the CAF 2025 flight plan (Pinter, 2004) and the missions modeled in RAND's KC-135 recapitalization AoA (see Kennedy et al., 2006).

These six missions are modeled as receiving tanker support from four refueling locations (see Figure 3.3). The first three sub-missions (air superiority, airborne electronic attack, and close controlled strike) are combined into one AR location. The remaining submissions (intra-theater strike, long-range strike, and large ISR support) occur at separate locations.

For the air superiority mission, the tanker supports a pair of two-ship F-22 CAPs. The mission is modeled as a continuous operation in which fighters and tankers cycle to maintain at least one two-ship of F-22s on CAP continuously. We parametrically varied the tanker-to-CAP distance and base–to–AR location distance. In some circumstances, fighters had enough fuel to remain on CAP after their replacement two-ship arrived. In these cases, we allowed the leaving two-ship to remain on station, providing additional coverage until it was at bingo fuel and had to either return to base or return the tanker.

The airborne electronic attack mission consists of four continuous unmanned combat aerial vehicle (UCAV) CAPs that conduct suppression of enemy air defenses, destruction of

Figure 3.3
Theater Employment Refueling Orbits

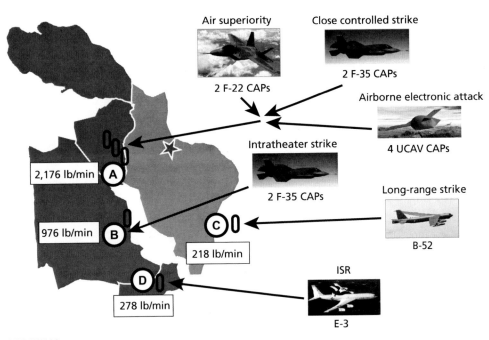

RAND *TR901-3.3*

enemy air defenses, and jamming. All four airborne electronic attack orbits are refueled from a single refueling point. We modeled the close controlled strike mission as two continuous F-35 orbits on call to ground forces to conduct air-to-ground strike missions.

The air superiority, airborne electronic attack, and close controlled strike missions are depicted in Figure 3.4; all are supported from a single AR area (Orbit A in Figure 3.3). For all missions in Orbit A, the tankers are based 750 nm from the AR location.

The intratheater strike mission is flown by a pair of F-35s that, after flying 400 nm to the AR location, hit a target 500 nm from the AR location and then return to base via the tanker for another AR. In the intratheater strike mission, the tanker is based 750 nm from the refueling location. The intratheater strike mission is serviced by AR location B, as shown in Figure 3.5.

In the long-range strike mission, shown in Figure 3.6, the tanker and B-52 bomber start from the same base and rendezvous for AR after flying 1,000 nm (halfway to the primary station). The B-52 continues on to strike targets, then returns to its originating base. The tanker refuels the B-52 at a rate of 5,500 lb per minute.

The last sub-mission of the theater employment mission is ISR support. In this case, the tanker travels 800 nm for AWACS refueling, offloading 95,000 lb at 5,500 lb per minute, as shown in Figure 3.7.

Deployment

The deployment mission consists of a tanker "dragging" a number of fighters from their home base in the United States to an overseas location across distances that the fighter aircraft could

Figure 3.4
Theater Employment Orbit A Mission Details

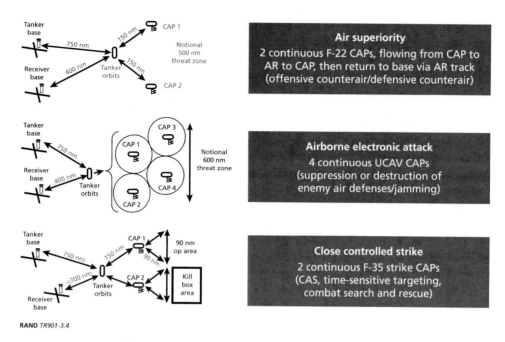

RAND TR901-3.4

Figure 3.5
Theater Employment Orbit B, Intratheater Strike Mission

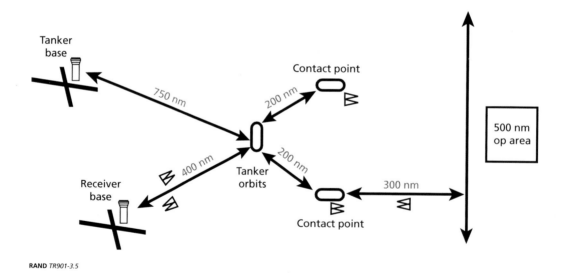

RAND TR901-3.5

not transit without AR support.[1] The deployment mission requires the continuous refueling of the fighter package during the deployment until the fighter package has enough fuel to reach its destination point without further refueling. Since the fighters must remain at a fuel state

---

[1]   *Dragging* is a term used colloquially in the U.S. Air Force tanker community to describe cases in which receiver aircraft fly in formation with the tanker as the lead. It occurs when tankers must refuel a number of fighter aircraft flying as a large formation over long distances (i.e., "fighter drag").

Figure 3.6
Theater Employment Orbit C, Long-Range Strike Mission

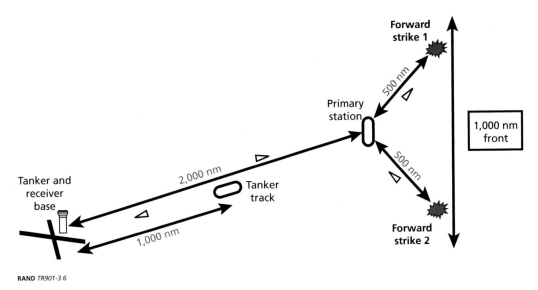

RAND *TR901-3.6*

Figure 3.7
Theater Employment Orbit D, ISR Support Mission

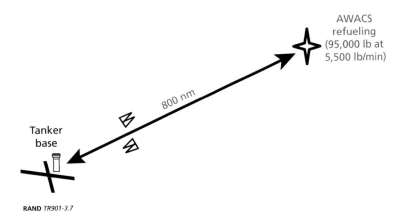

RAND *TR901-3.7*

that is high enough for them to reach a divert base in the event of a malfunction, each fighter must receive fuel regularly from the tanker. In the deployment mission analysis, we considered the deployment of three fighter types to five theater commands. The three fighter types (large, medium, and small) were considered representative of deployment missions in the 2024 time frame. We used the same fighter fuel demand rates that were computed in RAND's KC-135 recapitalization AoA (see Kennedy et al., 2006). The following fuel demands represent the average fuel usage of the tanker during the refueling portion of the mission (i.e., they include both tanker fuel burn and fuel offload to the receiver aircraft):

- small fighter: 295 lb per minute
- medium fighter: 830 lb per minute
- large fighter: 1,131 lb per minute.

Tanker Profile

In all cases, "round-robin" tanker missions were used for the deployment mission. Round-robin tanker missions allow multiple tankers to "relay" the fighter package during long deployment missions. Round-robin missions are often the most efficient use of tankers. In our analysis, the tanker profile consists of the tanker taking off, climbing to 30,000 ft, and flying 250 nm, where it will rendezvous with the fighter package (see Figure 3.8). The tanker then descends to 25,000 ft and drags the fighter package during the refueling segment. At the conclusion of the tanker's refueling segment, it returns to its originating base. If the required refueling distance is too far for one tanker, additional round-robin tankers are added. Another tanker picks up the fighters where the first tanker left the formation and drags the fighters for its refueling segment. This process is repeated until the fighters have enough fuel to reach their deployment location. In our model, all legs of the relay are equal for a given deployment distance, tanker configuration, and fighter package.[2]

Fighter Profile

In our analysis, the fighter profile used to support the tanker deployment mission analysis consists of the fighter package climbing to 25,000 ft and flying for 250 nm, at which point it meets up with a tanker. The package is then refueled until it is 1,500 nm from its destination point, where it leaves the tanker and flies the remaining distance without it.

Figure 3.9 shows the deployment missions evaluated in our study. We modeled deployment missions flying from bases in the United States to each of the five combatant commands (COCOMs) using the representative distances shown in the figure.

Figure 3.8
Tanker Deployment Mission

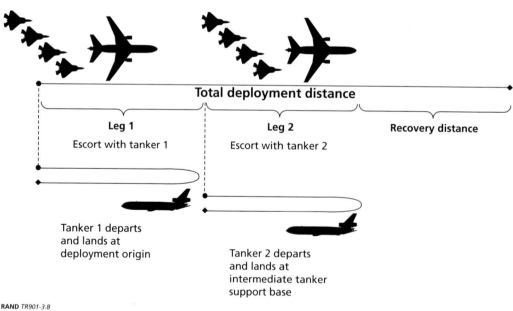

---

[2]   This approach is somewhat stylized in that it assumes that all round-robin tanker profiles are identical. In reality, each profile will be different due to the infrastructure available for the intermediate tanker base locations. However, it provides a reasonable generic modeling of the problem without assumptions about en route base availability.

Figure 3.9
Deployment Distances to Each COCOM

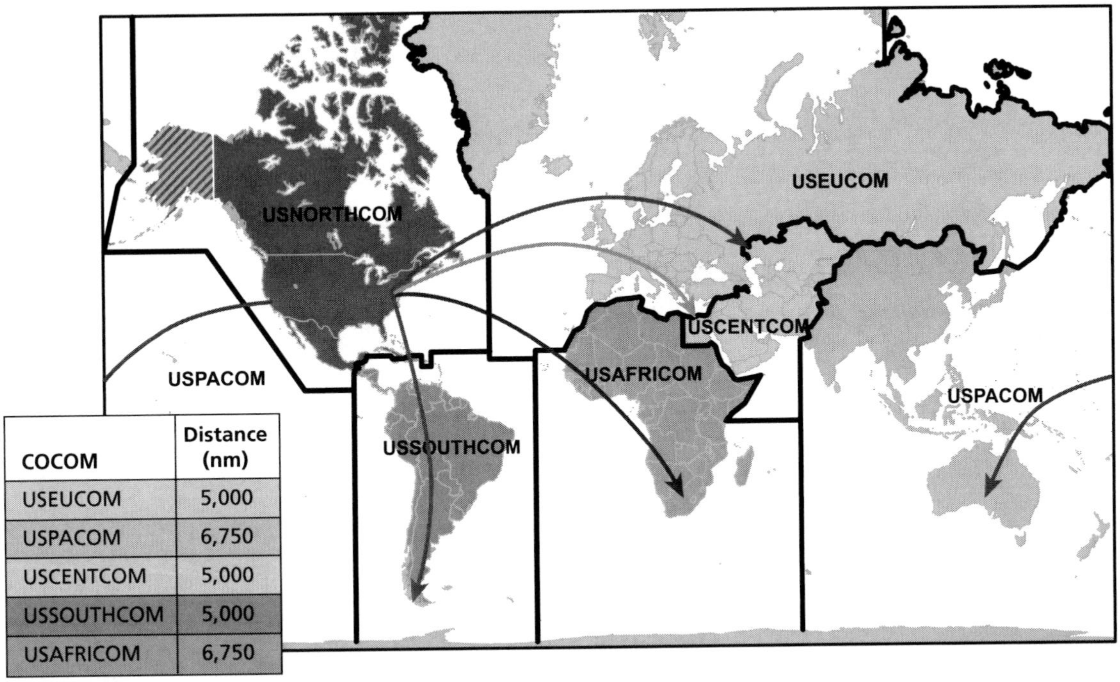

COCOM	Distance (nm)
USEUCOM	5,000
USPACOM	6,750
USCENTCOM	5,000
USSOUTHCOM	5,000
USAFRICOM	6,750

NOTE: USAFRICOM = U.S. Africa Command. USCENTCOM = U.S. Central Command. USEUCOM = U.S. European Command. USPACOM = U.S. Pacific Command. USSOUTHCOM = U.S. Southern Command.

RAND *TR901-3.9*

Air Bridge

The air bridge mission captures the requirement for tankers to refuel large aircraft receivers covering long distances to conduct their mission. By *large aircraft*, we mean aircraft that are larger than fighter aircraft and that do not require continuous refueling and escort to cross long distances. We use the mission title *air bridge* to capture the missions of global strike, air bridge, OPLAN 8010, and national reserve, as described in RAND's KC-135 recapitalization AoA (see Kennedy et al., 2006) and doctrine (specifically, Joint Publication 3-17 [JCS, 2009] and Air Force Doctrine Document 2-6, 2006). Although these four missions vary in their overall military purpose and goals, they are very similar from the perspective of the tanker operations required to support them. In these missions, large aircraft receive a single substantial offload from the tanker to extend their range, rather than requiring a continuous escort as in the deployment mission. We modeled air bridge missions with the tanker flying 1,000 nm to meet the receiver and then offloading at 5,500 lb per minute, retaining enough fuel to return to its originating base with reserves. This scenario is shown in Figure 3.10.

Missions, Modernization Areas, and Benefits

Each of the modernization options provides unique benefits for tanker missions and other warfighting missions. However, all modernization options do not provide benefits for every

Figure 3.10
Air Bridge Mission

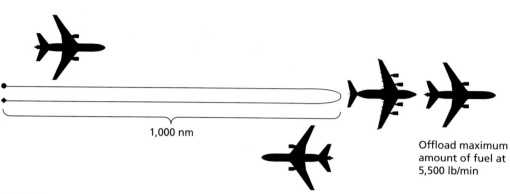

1,000 nm

Offload maximum
amount of fuel at
5,500 lb/min

mission, nor does each option provide every one of the listed benefits. Rather, each option uniquely benefits specific mission types. For example, defensive systems may allow tankers to base closer to AR locations and conduct AR closer to threats than without the systems. However, defensive systems do nothing to improve the rate at which receivers cycle across the boom or baskets. Additional multipoint capability would benefit the employment and deployment missions by allowing a formation of receivers to refuel in less time. Additionally, in the case of the deployment mission, an additional refueling point on a tanker may allow the tanker to escort and refuel a larger number of receivers in the cases in which the number of receivers is limited by their minimum fuel state.[3] Better C2 functionality may allow tanker aircraft to operate closer to fighter CAPs and enemy threats with intelligence that is more up to date. Better C2 information may also allow more efficient planning and operations by reducing the need for tanker aircraft to arrive at refueling locations ahead of their receiver aircraft. The benefit of eliminating this planned waiting time applies to all the mission areas. The primary benefit of NVIS-compatible lighting in the employment mission is in the potential for receiver aircraft to refuel faster because their pilots do not have to remove their NVG and allow their eyes to adjust before conducting AR operations. The benefits provided by each modernization option for each mission are shown in Tables 3.1 and 3.2.

KC-10 Mission Allocation

As discussed earlier, each modernization option provides unique benefits to different mission types. The dependence of benefits on the mission makes the relative weighting of the tanker missions important in determining the overall benefit of a given modernization option. For example, assume that modernization option A provided benefit only for mission area Z. If mission Z were the only mission that the KC-10 were required to conduct, then it is rather straightforward to account for the benefit from modernization option A. At the other extreme, if there were no requirement to execute mission Z, then there would be no benefit in choos-

---

[3]   During long deployment missions, each fighter aircraft must maintain a minimum fuel level to ensure that it can reach a divert location should there be a problem. In some circumstances, this minimum fuel requirement is so high that the fighters in the formation must continuously cycle across the boom so that the first fighter remains above the minimum fuel level after the last fighter finishes refueling.

Table 3.1
Modernization Options, Missions, and Tanker Efficiency Benefits

Modernization Option	AR Orbits Farther Forward	Tanker Bases Closer to AR Orbits	More Efficient Planning and Operations	Faster Receiver Cycle Times
TDL	Employment		Homeland defense Employment Deployment Air bridge	
Additional multipoint refueling capability				Employment Deployment
Defensive systems	Employment	Employment		
NVIS-compatible lighting				Employment

Table 3.2
Modernization Options, Missions, and Improvements to Other Military Missions

Modernization Option	Reduced Attrition	Increased CAF Range or CAP Coverage	Airlift Augmentation	Relay Augmentation
TDL	Employment	Employment		Employment
Additional multipoint refueling capability		Employment		
Defensive systems	Employment	Employment	Deployment	
NVIS-compatible lighting	Employment			

ing modernization option A. To explore the effect of mission distribution on the results of our analysis, we used two representative mission mixes. The first is an employment-heavy mix in which the majority of the tankers are dedicated to supporting a theater conflict. The second is a deployment-heavy mix in which almost two-thirds of the tankers are dedicated to supporting the deployment of fighter aircraft to the conflict. These two distributions are illustrated in Figure 3.11.

Summary

This chapter presented more detail on the wartime missions we used to model the effectiveness of the KC-10 modernization options. The missions were related to the expected benefits from the upgrades. The chapter also explained that effectiveness benefits are unique to each modernization option and that each option provides benefits to specific missions. Finally, we presented the two mission mixes used to evaluate mission effectiveness.

Figure 3.11
Employment- and Deployment-Heavy Mission Mixes

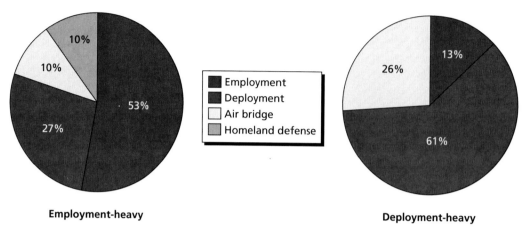

Employment-heavy   Deployment-heavy

RAND *TR901-3.11*

CHAPTER FOUR

Valuing KC-10 Modernization Options

In this chapter, we present background information and our methodology for our acquisition cost estimates for the four modernization options. All costs are in FY 2009 dollars. We then present our methodology for valuing the wartime effectiveness improvements that result from the upgrades. Finally, we present an approach to valuing improvements in KC-10 reliability, maintainability, and safety.

Upgrade Costs

We used two primary sources for our cost estimates. We were given access to KC-10 concept refinement studies conducted by five contractors under the KC-10 Aircraft Modernization Program. Some of these studies included cost information at a level of detail sufficient to attribute costs to specific components for some of the upgrades. We identify our derived estimates based on these studies as "contractor estimate A," "contractor estimate B," and so on, to avoid including proprietary information in this report. It is important to note that the designation contractor A, for example, does not necessarily refer to the same contractor in each of the following tables. Our second source of information for the cost estimates was P3A (procurement appropriation) exhibits in various years of the President's Budget (PB) documents on analogous programs.[1]

Table 4.1 shows our estimates of the cost per shipset for each of the four modernization options.[2] Flyaway cost is the cost of procuring and installing the equipment on the aircraft; average unit procurement cost (AUPC) is the sum of the flyaway cost plus ancillary items, such as initial spares, peculiar support equipment, training, and data. For most classes of equipment, we estimate AUPC to be 15 percent greater than flyaway cost, based on the average of the values in the contractor studies mentioned earlier. For some types of equipment, there was specific evidence of different AUPC–to–flyaway cost ratios; we note these cases where we discuss the cost estimates for those upgrades.

More detailed cost estimates for each of the modernization options are presented in Chapters Five through Eight.

---

[1] P3A (budget document) exhibits provide detailed budget histories and projections for modification programs. By assembling data from a series of these documents, it is possible to construct a set of annual data that can be converted to constant-year dollars. The FY 2009 PB was the latest available at the time of this study. Prior-year budgets were used to fill in earlier data. The years used for each modernization option are shown in the detailed cost estimates presented later in this report.

[2] By *shipset*, we mean the total amount of equipment to upgrade one tanker aircraft.

Table 4.1
KC-10 Modernization Cost Estimates

Modernization Option	Flyaway Cost per Shipset (FY 2009 $ millions)	AUPC per Shipset (FY 2009 $ millions)
TDL	0.5	0.7
Additional multipoint refueling capability	3.7	4.2
Defensive systems	15.1	19.2
NVIS-compatible lighting	3.1	3.6

Valuing Wartime Effectiveness

In this section, we discuss our methodology for assessing the value of improvements in KC-10 effectiveness. We express this in terms of the *percent increase* in effectiveness: If ten KC-10s, after modification, can provide the AR services that 11 can provide without modification, that is a 10-percent improvement. So far, we have presented our estimates of the cost of each modification option. With that information, and with the percent improvement that results from each modification option, we can rank the modification options by their cost-effectiveness. For example, say that modernization option i has cost C_i and percent improvement p_i. The ratio of percent improvement to cost is represented by the variable g_i. Thus,

$$g_i = p_i / C_i,$$

where g_i is a measure of the cost-effectiveness of each option. Dollars spent on the option with the highest g_i will deliver the most capability improvement per dollar, those with the second-highest g_i will deliver the secondmost capability improvement per dollar, and so on. If there were a fixed budget for KC-10 modernization, it should be spent on the options with the highest g_i.

While this approach tells us how options compare relative to each other (and how any fixed modernization budget should be spent across them), it does *not* tell us how large the budget should be, i.e., which options are worth implementing and which are not. The g_i allows us to rank the options in order of cost-effectiveness, but it does not tell us how far down the list to go. For that, we need a measure of *the value of improving AR effectiveness.*

The new KC-X program can be used to derive such a value. KC-Xs are being procured to provide a given level of AR services per aircraft at a given cost per aircraft. The program provides an indication of the value that the nation puts on the services: If the nation valued the services less, it would buy fewer KC-Xs; if it valued them more, it would buy more. Therefore, we use this approach to value the additional KC-10 AR capability that modernization could bring. If a potential KC-10 modification provides AR at a lower cost than a KC-X, modernization should occur; if it provides it at a higher cost, it should not.[3] Using this criterion ensures

---

[3] For this analysis, we assume that the marginal value of AR services is constant as their level changes, which is appropriate for the moderate size of the changes discussed here—single-digit changes in a small part of the entire AR fleet. KC-10s provide about 15 percent of total U.S. Air Force AR capability, according to RAND's KC-135 recapitalization AoA analysis (see Kennedy et al., 2006).

that the funding for KC-10 modernization provides at least as much AR capability per dollar as funding for KC-X does, which ensures that KC-10 modernization resources are being spent cost-effectively. As discussed earlier, the options assessed in this study have additional benefits besides those that we can translate into AR effectiveness. Thus, our criterion for valuing the modernization options is conservative in that it does not include those additional benefits. Modernization options that meet our cost-effectiveness criterion are robustly cost-effective, since we cannot include all the benefits in our calculation. In other words, if we did judgmentally assign a value to the additional benefits, the cost-effectiveness of the options would increase. We have chosen not to do so, which makes our findings conservative and robust, but others may wish to add such a valuation and thus modify our findings.

We based our estimate of the value of AR services on RAND's KC-135 recapitalization AoA (see Kennedy et al. 2006). The methodology of that analysis was intensively reviewed and vetted by both the Office of the Secretary of Defense for Program Analysis and Evaluation (now Cost Analysis and Program Evaluation) and the Institute for Defense Analyses.

The value of providing AR services with the KC-10 is based on two factors. The first is the cost of providing such services with a KC-X, and the second is the effectiveness of the KC-10 relative to the KC-X in providing the services. The KC-X competition was ongoing as of fall 2009. In an earlier competition, the Airbus 330 was chosen over the Boeing 767, but Boeing successfully protested this decision, and a new competition is expected soon. At this time, it is not known which contractors will compete for the program and with what airframes. Therefore, we use figures for a generic KC-X, based on RAND's AoA work. When a winner is chosen for the KC-X contract, these figures should be updated with the cost and effectiveness estimates for that specific aircraft. We judge the AoA results to be the best current basis for this study.

We found that the acquisition cost of the generic KC-X is $152 million, and the present value of the cost of operating it for 30 years is $160 million (all costs are expressed in FY 2009 dollars). We use a 30-year horizon because we assume that KC-10s will be operated until 2045 and that the modernizations will be installed by 2015, on average. Thus, the appropriate KC-X cost for comparison with the cost of KC-10 modernization is the cost of providing AR services with the KC-X for 30 years. In the RAND KC-135 AoA, KC-Xs were operated for 60 years, so we include only the part of the KC-X acquisition cost attributable to its first 30 years of operation. This is $105 million,[4] so the total cost of providing AR services for 30 years with the KC-X is $265 million, the sum of its acquisition and operating costs.

Also based on the AoA results, one KC-10 has 1.1 times the AR effectiveness of a KC-X. (The generic KC-X, like the Boeing 767 and the Airbus 330, is smaller than the KC-10 and so provides less AR capability per aircraft.) Therefore, the value of 30 years of AR services from a KC-10 is $290 million, and the value of a 1-percent improvement in KC-10 capability is $2.9 million.

As discussed earlier, we represent the cost of any modification as C_i and the resulting percent improvement in AR capability as p_i. Since the value of a 1-percent improvement in AR

---

[4] This figure is the result of converting the acquisition cost into a 60-year sinking fund whose present value equals the acquisition cost. The present value of the first 30 years of the sinking fund is our measure of the part of the KC-X acquisition cost attributable to its first 30 years of operation. This approach is appropriate if the effectiveness of the KC-X is constant throughout its 60-year lifetime, as it was in RAND's KC-135 AoA.

capability is $2.9 million, the value of a p_i-percent improvement is ($2.9 × p_i) million. Any modification is cost-effective if its value exceeds its cost; that is, if

$$(\$2.9 \times p_i) \text{ million} > C_i.$$

Valuing Improvements to Aircraft Reliability, Maintainability, and Safety

Figure 4.1 shows the history of the KC-10's depot-possessed rate, not-mission-capable (NMC) rate, and net unavailability rate from 1991 to 2008. The depot-possessed rate is the percentage of the total fleet that is in depot for heavy maintenance or modifications and not available for missions. The NMC rate represents the percentage of possessed aircraft (not in depot) that are unable to fly missions. The net unavailability rate is the sum of the number of aircraft that are NMC or depot-possessed divided by the total number of aircraft. Thus, the net unavailability rate is a comprehensive overall measure of reliability and maintainability: A higher net unavailability rate means poorer reliability and maintainability. Figure 4.1 shows that the net unavailability rate rose steadily from below 15 percent in 1991 to 35 percent in 2001 and then gradually fell to about 25 percent. Between 2007 and 2008, there was a sharp drop in the depot-possessed rate, but the NMC rate increased somewhat, so the net unavailability rate reflects only a modest decline.

Improvements in reliability and maintainability lead to increases in AR capability, just as capability modifications do. Requirements for AR translate into a requirement for *mission-capable aircraft*. Any improvement in reliability or maintainability will increase the number of mission-capable aircraft for a given TAI—that is, it will increase the net availability of the fleet,

Figure 4.1
History of KC-10 Depot-Possessed Rate, NMC Rate, and Net Unavailability Rate

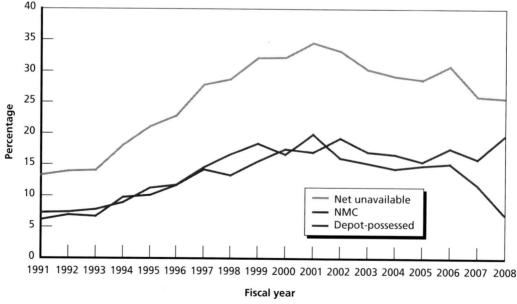

or the percentage of the fleet that is mission-capable. An increase in the number of mission-capable aircraft for a given fleet therefore increases AR capability, just as an increase in capability per aircraft does.

Improvements in reliability will decrease the number of repairs that must be made while aircraft are in depot and the rate at which command-possessed aircraft become NMC. (Command-possessed aircraft are the part of the TAI that is not depot-possessed.) Improvements in maintainability will decrease the time required per repair, both in depot and in the field, because they make NMC aircraft mission-capable. (We include under the general rubric of *maintainability* the supply of spare parts and the responsiveness of deliveries to needs for spare parts.) A simple formula relates the depot-possessed rate and the NMC rate to net availability:

a = net-availability rate
d = depot-possessed rate
n = NMC rate.

Therefore,

$$a = (1 - d) \times (1 - n).$$

With this formula, one can calculate how a change in either the depot-possessed rate or the NMC rate affects AR capability. This analysis uses the average depot-possessed and NMC rates over the 2004–2008 period: 13.5 percent and 17.0 percent, respectively.

Using the formula, we see that a 1-percent change in the depot-possessed rate will change the net availability rate by 0.83 percent. Thus, each KC-10 would be 0.0083 more mission-capable if the depot-possessed rate fell by 1 percent. AR capability increases by 0.0083 times the capability of one KC-10, which has a value of 0.0083 times $290 million, or $2.4 million. Any reliability or maintainability improvement package that increased the command-possessed rate by 1 percent would therefore have a value of $2.4 million per KC-10 and would be cost-effective if its cost were less than that.

Similarly, using the formula, we see that a 1-percent change in the NMC rate will change the net availability rate by 0.865 percent. Thus, each KC-10 would be 0.00865 more mission-capable if the NMC rate fell by one point. AR capability per KC-10 increases by 0.00865 times the capability of one KC-10, which has a value of 0.00865 times $290 million, or $2.5 million. Any reliability or maintainability improvement package that increased the NMC rate by 1 percent would therefore have a value of $2.5 million per KC-10 and would be cost-effective if its cost were less than that.

Any potential reliability and maintainability improvement should therefore be assessed in terms of both its cost and the resulting change in the depot-possessed rate and the NMC rate. Several such improvements have been proposed, including some in the contractor KC-10 Aircraft Modernization Program studies discussed earlier. However, none of these studies estimated how the improvements would affect the depot-possessed or NMC rates. Therefore, these improvements' cost-effectiveness cannot be assessed based on the information in the contractors' reports.

A detailed engineering analysis would be needed to estimate the impact of any reliability and maintainability modification on the depot-possessed or NMC rates. Such analyses were beyond the scope of our study but are necessary for making decisions about such modifications on a cost-effectiveness basis, as we are doing for effectiveness modifications. Making decisions about both types of modifications on a cost-effectiveness basis will ensure that overall modification funds are spent most cost-effectively—that is, that they achieve the greatest improvement in AR capability possible.

Using data from 2004 to 2008, we examined which components of the KC-10 were making the aircraft NMC to determine whether certain components were disproportionately responsible for this outcome and thus obvious candidates for reliability and maintainability improvement initiatives. There were none. Table 4.2 shows, for 2004–2008, how much of the total NMC rate was associated with the top five, top ten, top 25, and top 100 work unit codes (WUCs) and their impact as a percentage of the total NMC rate. The top five WUCs contributed only 14 percent of the total, and the top 100 were responsible for 47 percent. Therefore, reliability and maintainability improvements to many systems would be necessary to achieve a major improvement in net availability and reduce the NMC rate.

The contribution of individual WUCs to the total was not stable over time, which makes identification of specific systems to target for improving reliability more difficult. Table 4.3

Table 4.2
Contribution of Top WUCs to NMC Rate, 2004–2008

Number of WUCs	NMC Rate (%)	Percentage of Total NMC (17%)
Top 5	2.3	14
Top 10	3.2	19
Top 25	4.8	28
Top 100	8.1	47

Table 4.3
Top WUC Contributors to the NMC Rate, 2004–2008

WUC	Name	NMC Rate (%)
03724	ISO Inspection Maintenance Steering Group III 1A	1.1
13DAA	Wheel/Tire Assembly	0.4
23JC0	Thrust Reverser	0.3
46AA0	Integral Fuel Tanks	0.3
23000	Power Plant System	0.3
23002	Power Plant Tail	0.2
46AD0	Fuselage Bladder Tanks	0.2
04199	Special Inspection, Not Otherwise Coded	0.2
46ADC	Cell, Fuel Bladder, Forward #3	0.2
23003	Power Plants	0.2

NOTE: ISO = International Organization for Standardization.

shows the top ten contributors over the period 2004–2008 and how much of the total NMC rate was due to each.

Table 4.4 shows the rank of each of the ten WUCs by year from 2004 to 2008. The WUC that contributes most to NMC (WUC 03724, a time-phased inspection) was the top contributor in each year, but the other WUCs' positions were not stable over time.

Table 4.5 shows how much of the total NMC was due to each of the top ten WUCs in the years 2004–2008. The magnitudes of these contributions changed significantly over time.

We interpret this instability in the contribution of the top WUCs to the NMC rate as reinforcing our finding that there are no obvious approaches to most cost-effectively improve reliability and maintainability. Instead, detailed engineering analysis is needed to assess how much the NMC or depot-possessed rates might change with any specific modification. At

Table 4.4
Rank of Top WUC Contributors to the NMC Rate, by Year, 2004–2008

WUC	Name	2008	2007	2006	2005	2004
03724	ISO Inspection Maintenance Steering Group III 1A	1	1	1	1	1
13DAA	Wheel/Tire Assembly	4	2	2	2	3
23JC0	Thrust Reverser	2	5	4	5	7
46AA0	Integral Fuel Tanks	6	3	5	8	6
23000	Power Plant System	7	12	3	4	2
23002	Power Plant Tail	5	6	8	17	4
46AD0	Fuselage Bladder Tanks	8	4	7	11	140
04199	Special Inspection, Not Otherwise Coded	29	60	10	3	5
46ADC	Cell, Fuel Bladder, Forward #3	3	33	15	25	18
23003	Power Plants	16	13	9	7	9

Table 4.5
Contribution of Top WUCs to the NMC Rate, by Year, 2004–2008

WUC	Name	2008	2007	2006	2005	2004
03724	ISO Inspection Maintenance Steering Group III 1A	1.2	1.1	1.3	1.1	1.0
13DAA	Wheel/Tire Assembly	0.4	0.4	0.4	0.5	0.3
23JC0	Thrust Reverser	0.6	0.3	0.3	0.2	0.2
46AA0	Integral Fuel Tanks	0.3	0.4	0.2	0.2	0.2
23000	Power Plant System	0.3	0.2	0.3	0.2	0.3
23002	Power Plant Tail	0.3	0.2	0.2	0.1	0.3
46AD0	Fuselage Bladder Tanks	0.2	0.3	0.2	0.1	0.0
04199	Special Inspection, Not Otherwise Coded	0.1	0.1	0.2	0.3	0.2
46ADC	Cell, Fuel Bladder, Forward #3	0.4	0.1	0.2	0.1	0.1
23003	Power Plants	0.2	0.1	0.2	0.2	0.1

that point, changes in those rates, along with the cost of the modification and the value of improvements, as identified earlier, can all be combined to determine the cost-effectiveness of each particular modification. When attempting to determine changes in NMC and depot-possessed rates, care should be given to include the time that the depot possesses the aircraft to actually perform the modernization. Long modification times will serve to offset future availability improvements. In this study, we assumed that the modifications would be made during programmed depot maintenance cycles and that the programmed depot maintenance time would not be affected by this additional task. This assumption reflects our judgment that these specific modifications can be fit into the depot workflow without an overall increase in depot-possessed time for scheduled maintenance; however, this assumption is subject to further assessment.

This approach will ensure that capability modifications and reliability and maintainability modifications are valued on the same basis, which will lead to the most cost-effective use of modification funds.

We now turn to safety modifications—that is, modifications that will reduce the non-combat-related attrition rate of the KC-10. Since its introduction, the safety record of the KC-10 has been very good. Only one aircraft has been lost, in a 1987 ground accident; one life was lost in that accident.

To put a value on safety improvements, we must determine the cost of an aircraft loss, which is the sum of the cost of the physical aircraft loss itself and the cost of the lives that are lost. We refer to the cost of the loss of the physical aircraft as the *equipment-loss cost* and the cost of the loss of lives as the *loss-of-life cost*.

As an illustrative example of equipment-loss cost, say that a KC-10 is lost in year t, where t is some year between 2015 and 2045. (Here, we assume that safety modifications are made, on average, in 2015.) Had it not been lost in year t, it would have been operated for t additional years, where $t = 2,045 - t$. In our approach, if an aircraft is lost, its AR capability is replaced by an equivalent capability from the KC-X. We therefore consider the cost of adding those KC-Xs to the fleet in year t instead of in 2045, when they would replace the retiring KC-10s anyway. This is equal to the cost of acquiring and operating the KC-Xs for t years. We discussed earlier how we calculated this for $t = 30$; the same method applies for other durations. This cost is partially offset because the operating costs of the lost KC-10 are not incurred for t years. Figure 4.2 shows, for each year of loss from 2015 to 2045, the present value of (1) the cost of acquiring and operating 1.1 KC-Xs until 2045, (2) the cost of operating a KC-10 until then, and (3) the difference, which is the net equipment-loss cost of losing a KC-10.

Note that, in Figure 4.2, the cost of acquiring and operating 1.1 KC-Xs until 2045 is $290 million in 2015, which is consistent with our earlier results. If the KC-10 had a 0.1-percent annual attrition rate in each year from 2015 to 2045, the expected equipment-loss cost per KC-10 would be 0.001 times the net cost shown in Figure 4.2. The total equipment-loss cost associated with a 0.1-percent attrition rate is the present value of all annual costs— $1.4 million. (A small adjustment must be made for the fact that the expected size of the fleet falls over time due to the attrition rate; the difference in the result is less than 1 percent.) This is the value of reducing the attrition rate by 0.1 percent, not including loss-of-life costs.

Valuing loss of life is a difficult issue because of the personal tragedy associated with it. Nonetheless, some valuation must be used. The measure we use here is the compensation that was given for victims of the September 11, 2001, attacks: $6 million per life. In this analysis, we assume that four lives are lost in missions that do not carry passengers, that 18 lives

Figure 4.2
Present Value of the Cost of Acquiring and Operating 1.1 KC-Xs Until 2045, the Cost of Operating a KC-10 Until 2045, and Their Difference

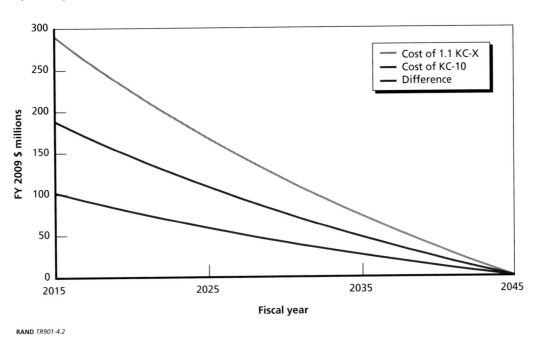

are lost in missions that do carry passengers, and that passengers are carried in 25 percent of flight hours. Therefore, 7.5 lives are lost per aircraft lost, at a cost of $45 million; the expected loss per year per KC-10 at an attrition rate of 0.1 percent is $450,000. The total loss-of-life cost associated with a 0.1-percent attrition rate is the present value of all the annual costs, or $1.4 million. (In this case, we also made the small fleet-related adjustment mentioned earlier.) This figure can be directly adjusted for different loss-of-life valuation (i.e., if the value of a life were judged to be $12 million, the cost would double).

The total value of reducing the attrition rate by 0.1 percent is the sum of the equipment-loss cost and the loss-of-life cost associated with the attrition rate, or $2.8 million. However, given the high value of one aircraft, preventing a single aircraft loss (even before accounting for the crew) would offset a significant portion of the upgrade cost. As with reliability and maintainability modifications, a detailed engineering analysis would be needed to estimate the impact of any safety modification on the attrition rate. Once that were done, it would be possible to compare the cost of the modification with its value, where the value is $2.8 million times the change in attrition (in tenths of a percent). Such engineering analyses were beyond the scope of this study but are necessary for making decisions about safety modifications on a cost-effectiveness basis. Decisions made on the basis of effectiveness, reliability and maintainability, and safety modifications, as we have illustrated here, will ensure that overall modification funds are spent most cost-effectively—that is, that they achieve the greatest improvement in AR capability possible.

Summary

In this chapter, we presented our methodology and overall acquisition cost estimates for the four modernization options. More detailed estimates and changes in peacetime operating costs associated with the upgrades are discussed in Chapters Five through Eight. This chapter also outlined our methodology for valuing the wartime effectiveness improvements that could result from the upgrades. Finally, we presented an approach to valuing improvements in KC-10 reliability, maintainability, and safety that would ensure that those investments and effectiveness improvements were as cost-effective as possible.

CHAPTER FIVE

Tactical Data Link

In this chapter, we present our analysis of improving the KC-10's C4 capability by adding a TDL. We provide some background information, followed by cost estimates for the upgrade, changes to wartime effectiveness, the implications of moving AR orbits in theater closer to operations, and the net present value (NPV) of adding the TDL. This modernization option is robustly cost-effective with a positive NPV.

Background

In the *Initial Capabilities Document for Aerial Refueling* (USAF, 2005c), *C4 for AR* refers broadly to improved connectivity with other aircraft as well as Air Force and joint information networks. In addition to standard military aircraft radios, the KC-10 has limited connectivity through a satellite Iridium phone that is connected to the intercom system. In this study, we restricted our analysis to the installation of a TDL on the KC-10. A TDL would allow secure data and voice communication between the KC-10 and receiver aircraft, air operations centers, or any other similarly equipped aircraft, as well as ground centers, ships, or vehicles. Although we focus exclusively on a TDL in this chapter, we explore how this information would be beneficial to the warfighter and, specifically, how it could make the KC-10 more effective in conducting its wartime missions. To that end, we examine how better information sharing may reduce or eliminate the additional flying time required for the tanker to arrive early on station for AR operations and how data links could allow tankers to facilitate more efficient rendezvous. For this effort, we estimated the cost of installing a Link 16 system on the KC-10 and the resulting operational benefits.

The current U.S. Air Force KC-10 fleet relies on analog voice communication to exchange information with air traffic control centers, C2 authorities, and other aircraft. The need for a civil digital data link to supplement voice communication and meet forthcoming CNS/ATM mandates was addressed in *Assessing the Cost-Effectiveness of Modernizing the KC-10 to Meet Global Air Traffic Management Mandates* (Rosello et al., 2009). There are additional benefits from adding a military TDL as well, including enhanced situational awareness for the tanker crew, improved TDL network coverage for other aircraft, and reachback capability for other network users.

The concept of network-centric operations has been a driving force behind many recent force modernization efforts in the U.S. military and some coalition militaries. Increased interoperability between systems and improved information sharing may lead to more efficient and effective combat operations. One key component of network-centric operations is the TDL, which allows users to exchange formatted digital messages over a radio network (Wilson, 2007).

Several TDL systems have been developed over the years to meet the needs of particular user communities. Some examples of legacy systems are Link 4, which provides C2 for fighter aircraft, and Link 11, which is employed primarily to share sensor data. Link 16, the most recent system, is designed to take over all of these functions and provide additional flexibility and increased data rates. It is a secure, jam-resistant data link based on the time-division multiple-access method for frequency sharing within a network. Link 16 also incorporates features not available on previous systems, including two channels for secure voice communication, along with navigation and flexible network capabilities.

A limited number of active network participants transmit and receive Tactical Digital Information Link J (TADIL-J)–series messages over the Multifunctional Information Distribution System (MIDS) or Joint Tactical Information Distribution System (JTIDS) ultrahigh frequency (UHF) radio terminals. An unlimited number of properly equipped passive participants can be accommodated as well. While UHF transmissions require a LOS between network participants, a Link 16 user equipped with a range-extension system can forward information to a beyond-LOS (BLOS) terminal. In addition, a gateway system can be used to translate between Link 16 and another TDL protocol, such as SADL, which is used by CAS aircraft. There is currently a deployed operational system called Joint Range Extension (JRE) that serves both of these roles.

Upgrade Costs

Table 5.1 shows two contractor estimates of the flyaway shipset cost for installing Link 16 on the KC-10 and the budget cost of procuring and installing Link 16 on four other aircraft. The two contractor estimates are very close, at $0.5 million.[1] The four budget costs are quite diffuse, with an average of just over $1 million. However, this average is pulled up by the very high B-1B program cost. We judge that the contractor cost estimates are reasonable in light of the budget costs, and we used the average of the two as our estimate. (We note here that our

Table 5.1
TDL Upgrade Cost Estimate Sources

Source of Cost Estimate	Flyaway Cost per Shipset (FY 2009 $ millions)
Contractor A	0.50
Contractor B	0.51
AC-130 (PB 2005–2009)	1.01
B-1B (PB 2003–2006)	2.74
F-15E (PB 2003)	0.24
F-16 (PB 2004–2009)	0.47

---

[1] We show contractor estimates as "Contractor A," "Contractor B," and so on, for each modernization option. These designations do not necessarily refer to the same contractor in different tables.

finding that this upgrade is cost-effective is very robust; even at the B-1B level, it would still be cost-effective.) We applied a 35-percent flyaway-to-AUPC markup to this estimate, based on the markup in the AC-130 program, resulting in an AUPC per shipset of $0.7 million

We assume that the inclusion of a TDL on the KC-10 would not increase network management cost, nor would an additional crew member be required to operate the system.

Valuing Wartime Effectiveness

The primary measure used to determine the potential value of adding a TDL to the KC-10 was the percentage increase in tanker effectiveness. By reducing the tankers' required lead time and expediting rendezvous, a TDL improves tanker effectiveness in all missions.

Current operational practice is for tanker aircraft to arrive at AR locations 15 minutes prior to scheduled refueling operations. This overlap is specified in air tasking orders from the current operations in Iraq and Afghanistan and is scheduling practice in the continental United States. Having the tanker arrive early reduces the chance that a receiver aircraft will be forced to divert because of a delayed tanker, but this margin comes at the cost of additional tanker resources. Tankers with TDLs would have real-time information on the location and status of receivers, allowing the tanker to time its departure and arrival with more accuracy. Figure 5.1 shows how the relative effectiveness of a KC-10 changes in various mission types as the preplanned time margin varies from the 15-minute nominal all the way down to none.

In addition to increasing efficiency by reducing tankers' required lead time, a TDL would be able to expedite every rendezvous. If both the tanker and receiver had real-time position awareness of the other, the tanker could ensure that it was in the optimal location in the prescribed refueling airspace to meet up with the receiver aircraft. To illustrate this expedited

Figure 5.1
Change in Effectiveness from Reduced Tanker Coverage Overlap

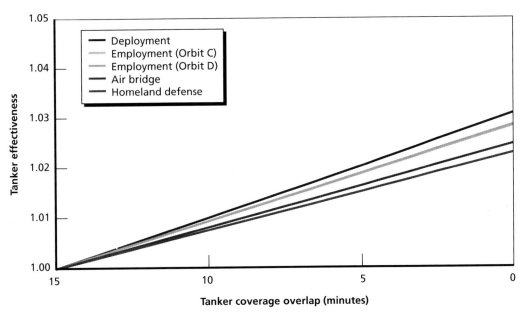

scenario, we use an example of a point parallel rendezvous. While not all AR operations are conducted as point parallel, the concept is similar in that there is a designated airspace and time in which the meet up will occur; if the tanker is optimally positioned within that airspace, the rendezvous time can be decreased.

A typical point parallel procedure requires a minimum track of 70 nm between the AR initiation point and the AR control point. The tanker orbits around the control point until the receiver arrives at the initiation point and gives radio notification. The tanker then turns to the reciprocal of the receiver's inbound track and continues until it reaches the appropriate turn range obtained from a table lookup. The final turn back into the inbound track is then initiated, positioning the tanker 1 nm in front of the receiver. If no radio contact has been established before the designated AR control time, the tanker must remain on station for an additional 10 minutes past the control time (NATO, 2010). The efficiency of this process, depicted in Figure 5.2, depends on how far along the track the rendezvous occurs. This varies based on the location of the tanker in its orbit when the receiver arrives at the AR initiation point, as shown in Figure 5.3.

A TDL provides additional situational awareness to the tanker crew, allowing them to track receiver aircraft. The result is an enhanced planning capability that could allow for reduced or eliminated tanker coverage overlaps and more efficient rendezvous.

Other Operational Benefits

Current tanker aircraft lack the survivability for missions near enemy threats and are typically operated far from these threats until air superiority is achieved. While a drastic departure from this doctrine is unlikely, the addition of defensive systems and improved situational awareness from a TDL might allow tanker AR orbits to be safely moved some distance forward earlier in

Figure 5.2
Point Parallel Rendezvous

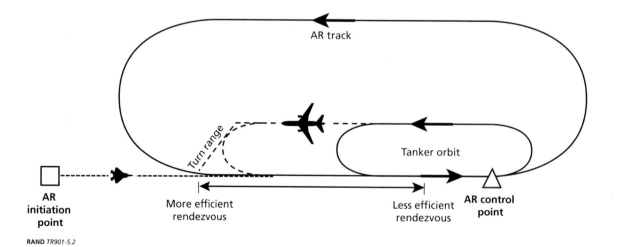

RAND TR901-5.2

Figure 5.3
Change in Effectiveness Due to Rendezvous Distance Improvement

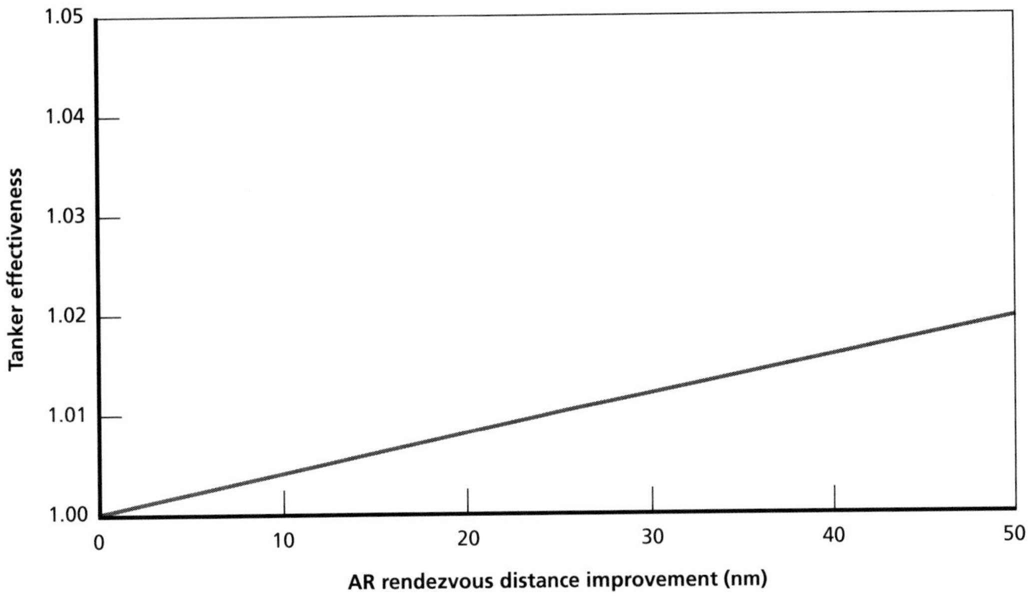

NOTE: Based on employment orbits A and B.

RAND *TR901-5.3*

the battle without increased vulnerability. This has the potential to improve the overall effectiveness of the warfighter in employment missions.[2]

Closer AR orbits would reduce the distance that fighter aircraft are required to travel to refuel between CAP shifts. This would allow fighters to stay on station longer before refueling is required, increasing CAP coverage overlap. Alternatively, CAPs could be pushed farther forward with the same level of coverage overlap. Strike aircraft could also penetrate deeper in this scenario, potentially increasing the number of targets accessible in a given combat radius.

These additional benefits come at the cost of increased workload on the tanker. Figure 5.4 shows a conceptual picture of a single AR tanker orbit supporting one fighter CAP. The tanker effectiveness and average fighter CAP coverage overlap for this nominal configuration are also shown in the figure.

As the AR orbit is moved forward toward the fighter CAP, both the tanker and fighter must travel greater distances from their bases to reach the deeper AR orbit location. The increased fuel burn of the tanker reduces the fuel available for offload, and the additional fuel required by the fighter increases the receiver offload demand. The combined effect results in a decrease in tanker effectiveness, as more tankers are required to support these missions. On the other hand, the reduced AR-to-CAP distance allows fighters to spend more time on station before refueling, increasing the CAP coverage overlap. Figure 5.5 illustrates the trade-off

---

[2]   Based on an examination of historical tanker operations; current operational tactics, techniques, and procedures; doctrine; present and emerging threats; situational awareness tools and equipment, we judged that conducting AR operations closer to threats than current practice was unlikely, even with the addition of a TDL and defensive systems. Nonetheless, we present a parametric analysis highlighting the implications for tanker effectiveness should AR operations occur farther forward.

Figure 5.4
Baseline AR Orbit Location

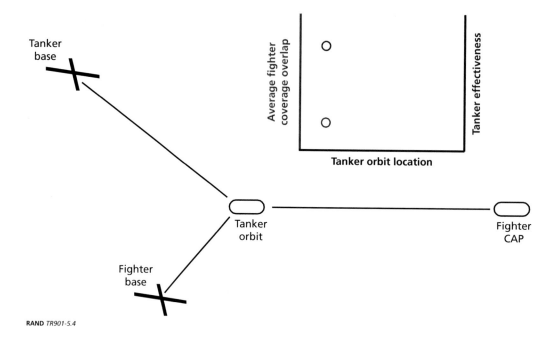

RAND *TR901-5.4*

Figure 5.5
Effect of Moving Tanker Orbits Closer to CAPs

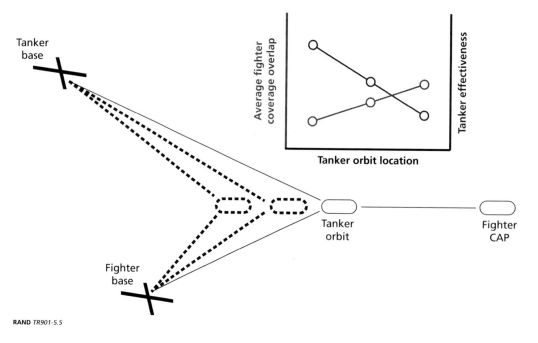

RAND *TR901-5.5*

between tanker effectiveness and fighter CAP coverage overlap that results from moving the tanker orbit forward.

Figure 5.6 shows the variation in tanker effectiveness and average fighter CAP coverage overlap with tanker orbit location for the most demanding tanker orbit in the employment

Figure 5.6
Effect of Moving Tanker Orbits Closer to CAPs on Employment Orbit A

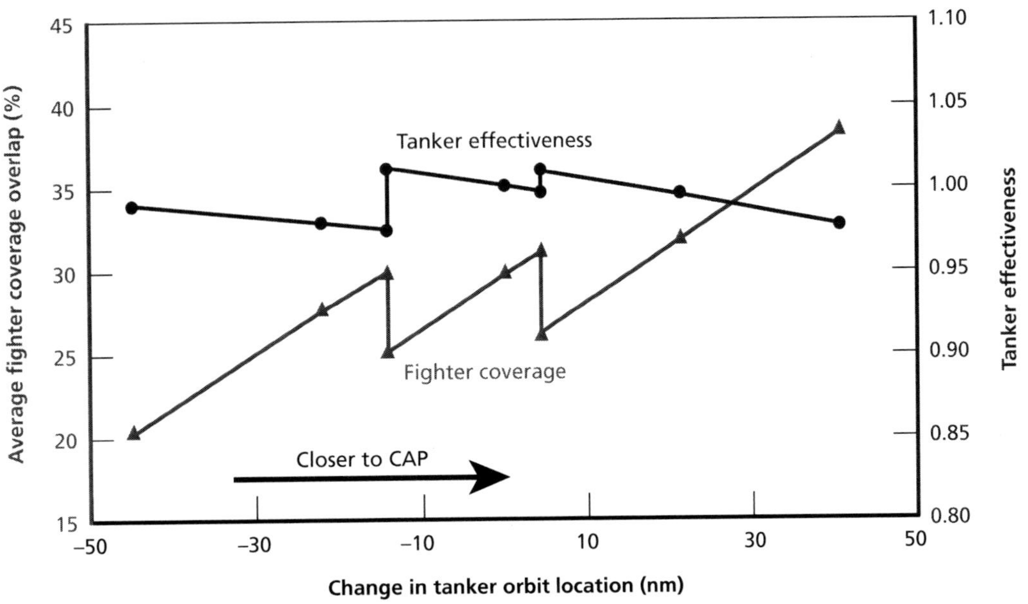

NOTE: Based on employment orbit A.

mission, as described in Chapter Three. It should be noted that as the tanker orbit is moved closer to the CAPs, the round-trip transit time for the fighters to refuel between CAP rotations decreases. This transit time determines the minimum number of fighters required to maintain a constant CAP presence. The discrete jumps in the curves represent transitions in the required number of two-ships from three to two for some subset of the modeled employment missions. At these points, the average refueling demand on the tanker decreases due to the smaller number of fighters that require support. As a result, there are discrete increases in tanker effectiveness and decreases in CAP coverage overlap.

The additional situational awareness provided by a TDL could also prevent the damage or loss of tanker aircraft. Although the attrition rate for tankers is extremely low, the potential for midair collisions and pop-up infrared (IR) threats still exists. As unmanned aerial vehicles continue to become more prevalent, the risk of collisions will likely increase. The proliferation of man-portable air defense systems (MANPADS) also presents a growing threat to large aircraft like the KC-10. A data link, such as Link 16, allows the sharing of threat warnings, air tracks, and other data that can help counteract these risks.

The persistence of tankers near combat operations would allow them to serve as LOS links between geographically separated network users. The normal range of a Link 16 message transmission is 300 nm, but access to a network can also be limited by terrain. A tanker could potentially provide a LOS link between aircraft flying at low altitudes near terrain while still serving in its primary AR role. Figure 5.7 illustrates the potential coverage improvement provided by a single tanker flying at 25,000 ft above mean sea level and serving as a LOS link to a CAS aircraft flying at 2,000 ft above ground level.

Equipage with a system such as JRE would also allow tankers to translate data and voice messages between different TDL networks, increasing the amount of information available to

Figure 5.7
Terrain-Limited Network and Coverage Improvements from a KC-10 LOS Link

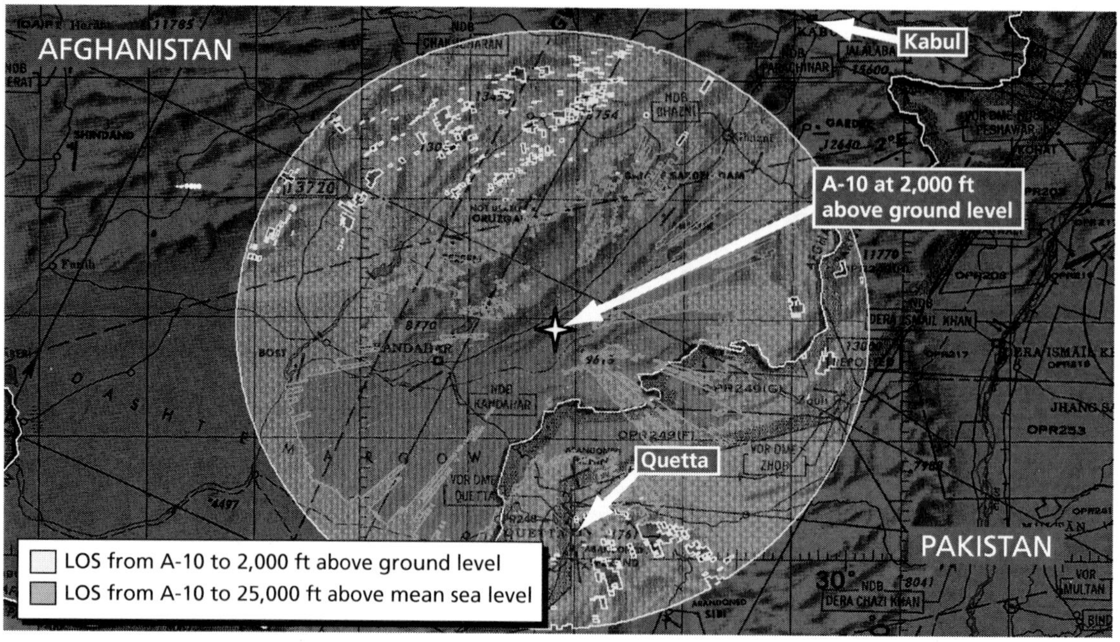

RAND *TR901-5.7*

users of different systems. The JRE can forward data BLOS over STU-III (secure telephone unit), SIPRNet (Secret Internet Protocol Router Network), satellite, or other secure means, creating a global network that links operational theaters around the world (Camana, 2008).

Tactical Data Link Cost-Benefit Summary

In this section, we summarize the costs and calculated benefits of upgrading the KC-10 with a TDL. All costs and benefits are in terms of NPV. NPV calculations take into account the time value of money and should be thought of as incorporating associated costs and benefits over the remaining life of the aircraft.[3]

The baseline assumptions used in the benefit calculations include a 5 nm reduction in the distance required for tanker-receiver rendezvous and the elimination of the 15-minute planned overlap for tankers on station. These assumptions are based on the improvements in planning and operational efficiency expected to result from the increased situational awareness provided by fleetwide TDL equipage. We chose 5 nm for the rendezvous improvement based on the average airspace dimensions designated for AR in Operation Iraqi Freedom (OIF) and Operation Enduring Freedom. Given the length of the designated AR areas, 5 nm represents a reasonable improvement. The 15-minute overlap time was reduced to zero for these calculations because, in our judgment, the tanker crew's increased awareness of receivers' position and status would offset the need to arrive early and wait for 15 minutes. Tanker AR orbits were not

---

[3]   NPV is the appropriate way to judge modernization investments based on the true resource cost. The Office of Management and Budget (OMB) directs this kind of discounted analysis. We used the current (December 2009) OMB-directed real long-term discount rate of 2.7 percent in our analysis, which represents the "return on investment" (OMB, 2009).

moved forward in the employment missions for the baseline analysis, since our judgment is that the additional risk incurred by the tankers would not justify the benefits in fighter coverage and penetration. Table 5.2 shows the effectiveness changes that result from these operational benefits. Note that the effectiveness for the employment mission is a weighted average of the effectiveness of the four orbits, A–D, as shown in Figure 3.3; orbit A is weighted four times as much as the others because of its much larger demand.

To determine the resulting NPV of the benefit, these effectiveness changes were valued using the tanker missions and mission mixes described in Chapter Three. Table 5.3 shows the cost-benefit summary for adding TDL to the KC-10. The TDL is robustly cost-effective because it is relatively inexpensive compared with the other options and provides benefits in all missions. It is cost-effective even with zero improvement in the employment missions and as little as two minutes' time reduction in the other missions.

Table 5.2
Effectiveness Changes Resulting from Fleetwide TDL Equipage

Mission Type	KC-10 Wartime Effectiveness Δ (%)
Homeland defense	2.3
Employment	1.0
Deployment	3.1
Air bridge	2.5

Table 5.3
TDL Cost-Benefit Summary

AUPC (FY 2009 $ millions)	Mission Mix	NPV Benefit (FY 2009 $ millions)
0.7	Employment-heavy	5.3
0.7	Deployment-heavy	7.7
0.7	Average benefit	6.5

CHAPTER SIX

Additional Multipoint Refueling Capability

Background

Tanker aircraft that have multipoint capability can refuel two fighter-size aircraft simultaneously. This is accomplished through the use of pods that are attached to the outboard sections of the wings of the tanker. Currently, this capability is limited to probe and drogue refueling, but "wing booms" are potentially viable.[1] In this analysis, we focus on the potential added effectiveness of multipoint refueling of probe-equipped aircraft only. However, the methodology we employ for determining the cost-effectiveness of probe and drogue refueling could just as easily be applied to evaluate the benefit of developing a capability for boom-receptacle multipoint refueling.

Only a fraction of the KC-10 fleet is currently capable of multipoint refueling. Twenty of the 59 KC-10s are equipped with the necessary fuel plumbing to accommodate wing-mounted refueling pods. The Air Force has 15 pairs of the refueling pods, but there are reliability problems with the current KC-10 refueling pods, which greatly limits their operational utility. In this analysis, we evaluate replacing all the refueling pods with a new system and plumbing the remaining 39 aircraft.

Multipoint refueling capability allows a set of receivers to refuel in less time because of the availability of an additional refueling point. Additional multipoint capability has the potential to improve operational effectiveness during both deployment and employment operations. It also allows a set of receivers to complete refueling faster, which can lead to more efficient tanker operations, thereby allowing a single tanker to refuel more aircraft than would be possible with single-point refueling. In addition, since a set of receivers requires less total time for refueling, the first receiver to refuel will have more fuel when the last receiver in the package finishes refueling. This additional fuel could provide more time for the receivers to conduct missions, such as by increasing time on station for air defense aircraft allowing deeper strikes for strike aircraft. The ability to refuel a package faster could also permit larger fighter packages per tanker.[2]

---

[1] Wing booms have been discussed for nearly 40 years, but only limited engineering studies have been conducted, and significant technical issues remain. The results of one such study are reported in Boeing Company, 1972. See also Kalt, 2004.

[2] The maximum time on station or maximum range of a strike package is often determined by fuel level. That is, the fighter package must return to the tanker when one of the fighters reaches a critical level of fuel. Multipoint refueling allows all the fighters to leave the tanker at a higher fuel state, since less time is spent waiting for subsequent fighters to be refueled.

Upgrade Costs

Table 6.1 shows the budget cost of procuring and installing the plumbing and refueling pods in both the KC-10 and the KC-135. (As noted earlier, 20 of the KC-10s have already been retrofitted with multipoint capability.) We used the KC-10 budget cost as our estimate. We applied the standard 15-percent flyaway-to-AUPC markup, resulting in an AUPC per shipset of $4.2 million.

Valuing Wartime Effectiveness

The primary measure used to determine the potential value of multipoint refueling was the percentage increase in tanker effectiveness that could be achieved by adding this capability to the KC-10 fleet. This analysis extended the methodology developed during RAND's KC-135 recapitalization AoA (Kennedy et al., 2006). Two tanker missions evaluated in that study benefited from the additional multipoint refueling capability: deployment and employment.

Deployment Mission

The deployment mission consists of a tanker dragging a number of fighters from their home base in the United States to an overseas location, as described in Chapter Three. An important constraint on the potential effectiveness of multipoint refueling in this mission is the limit on the number of fighters that can be dragged by a single tanker. International standards for AR state that, at night or in instrument weather conditions, no more than 12 aircraft should normally be in formation with the tanker, and the number of receivers in close formation should be limited to six (three on each wing) (NATO, 2010). Three per wing would limit the total number per tanker to six or 12. The rationale behind the maximum is that that three per side is a reasonable number to ensure aircraft separation and avoid collision through procedural means if the formation inadvertently enters clouds and the formation members lose sight of each other. There is uncertainty and difference of opinion with regard to the maximum number of fighters per tanker, so we looked at two possibilities. The first was to constrain the number of fighters per refueling point at six. In this case, we were able to drag 12 fighters in the multipoint case versus six in the single-point case. The second was to constrain the number of fighters to a maximum of eight per tanker. In this case, only eight fighters were dragged in the multipoint case versus six in the single-point case. Depending on the characteristics of the scenario (e.g., total offload fuel required, maximum refueling rate, deploy-

Table 6.1
Additional Multipoint Refueling Capability Cost Estimate Sources

Source of Cost Estimate	Flyaway Cost per Shipset (FY 2009 $ millions)
KC-10 Plumbing (PB 1996)	1.32
KC-10 Pods (PB 1996)	2.38
KC-135 Plumbing (PB 1999–2004)	3.03
KC-135 Pods (PB 1999–2004)	2.14

ment distance), the maximum number of fighters per tanker can limit the potential effectiveness increase obtainable from multipoint refueling.

Analytical Methodology. We measured the increase in effectiveness from multipoint refueling by determining the effectiveness ratio of multipoint verses single-point refueling. The effectiveness ratio is the total number of single-point tankers divided by the total number of multipoint tankers required to conduct the specific refueling task. For example, a refueling task could be dragging 48 fighters. This requires eight packages of six fighters or, in the case of multipoint refueling, six packages of eight fighters. The equivalency ratio includes the total number of tankers required to accomplish the fighter drag, accounting for ground turn times. This approach provides a measure of the total tanker time that is devoted to conducting the operation and gives the difference in effectiveness of the various options.

The round-robin analysis results in a step function for the number of tankers required. One tanker is able to drag a number of fighters up to a certain distance and then a second round-robin tanker must be added beyond that distance. As the deployment range is further increased, a third round-robin tanker must then be added. Since the fuel offload and tanker burn rate is different for the single-point and multipoint cases (due to the increased drag of the refueling pods), these steps occur at different deployment ranges in the two scenarios. We refer to these discrete steps as *integer effects*. The relative effectiveness (multipoint/single-point) changes as a result of the integer effects. For example, below a specific deployment range, a package of eight fighters using multipoint refueling and a package of six fighters using single-point refueling may both require only one tanker. As the range is increased, a second tanker will be required to deploy the eight-fighter package, while the six-fighter package will still require one tanker. In this case, two tankers flying equal round-robin profiles are needed to deploy the eight-fighter package, while only a single-tanker sortie is required for the six-fighter package. As the deployment distance is further increased, we reach a point where two tankers are required to deploy the six-fighter package. In that case, we then have two round-robin tanker profiles to deploy both packages. This simple example shows how the steps occur at different deployment ranges for different offload cases. Figures 6.1, 6.2, and 6.3 show the number of round-robin tanker legs required to deploy large, medium, and small fighters, respectively, as a function of deployment distance.

In all cases, we deployed fighters as a squadron of 24 aircraft because it allowed a fair comparison between the options and isolated the effects of multipoint refueling. We used 24 because it results in an integer number of deployments for all the fighter package sizes we considered (four, six, eight, and 12). That is, deploying 24 aircraft, four at a time, requires six packages, while deploying 12 fighters at a time requires two packages. We first computed the number of tankers required to deploy each package over the required distance. Larger packages generally require more tankers, but fewer packages deliver the entire squadron (except for very short distances, in which case the smaller packages have unused tanker capability). We then multiplied the number of tankers required to deploy each package size by the number of packages required to deploy 24 aircraft. For example, in the case of the eight-aircraft package size, three packages are required. In all cases, we assumed that a sufficient number of tankers would be available to deploy all fighters simultaneously. This is a bit of an analytical simplification and is not likely in operational practice, since there would likely be temporal separation between packages. However, by taking this approach, differences in time of closure for the squadron are eliminated as an independent variable, allowing us to calculate the effect of multipoint refueling capability on tanker effectiveness.

Figure 6.1
Number of Round-Robin Legs, Large Fighter Deployment

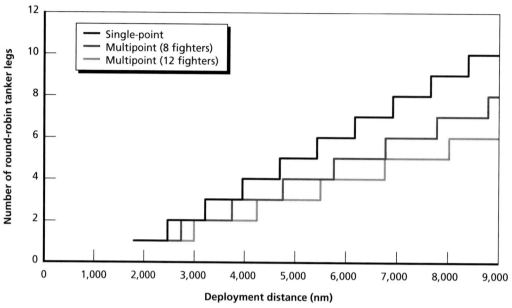

RAND *TR901-6.1*

Figure 6.2
Number of Round-Robin Legs, Medium Fighter Deployment

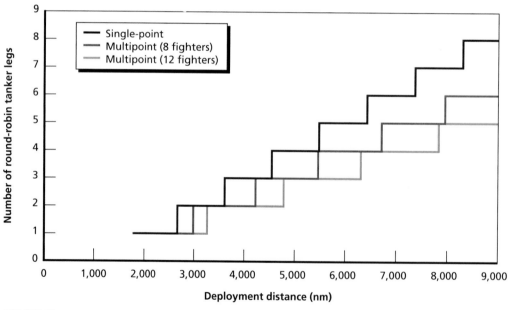

RAND *TR901-6.2*

Once the tanker profile is known and the number of round-robin tanker legs has been determined, we can compute the tanker flight time and ground time for each tanker sortie. Using these times, we determined the number of tanker aircraft that must be allocated to deploy the various packages to different ranges. We then determined the effectiveness ratio of

Figure 6.3
Number of Round-Robin Legs, Small Fighter Deployment

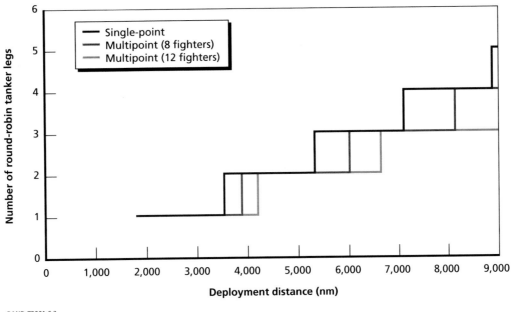

multipoint relative to single-point refueling operations. Figures 6.4 and 6.5 show the relative effectiveness of multipoint versus single-point refueling for the eight-fighter and 12-fighter multipoint packages. The figures also show the average effectiveness at each range. The average was computed based on the following assumptions regarding the breakdown of the type of aircraft supported in the deployment mission: 15 percent small, 34 percent medium, and

Figure 6.4
Relative Effectiveness of an Eight-Fighter Multipoint Versus Single-Point Package

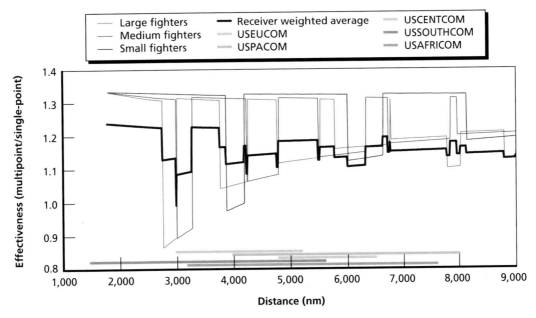

Figure 6.5
Relative Effectiveness of a 12-Fighter Multipoint Versus Single-Point Package

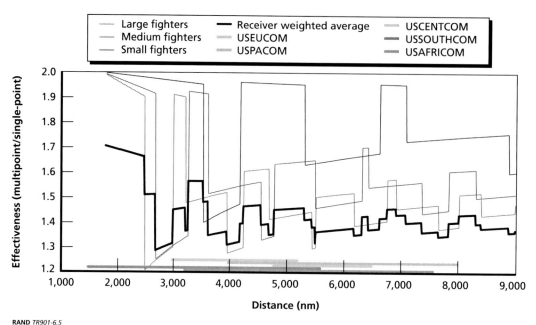

RAND *TR901-6.5*

22 percent large receivers. The remaining 29 percent would be heavy receivers for which there is no multipoint option, so this was assumed to be 1.0.[3] The solid bars near the bottom of the figures represent the range of potential deployment distances for each of the COCOMs.

Deployment Results. Finally, we computed the average effectiveness ratio (multipoint/single-point) for each of the COCOMs. The averages were uniformly weighted across the entire distance of the COCOM. Table 6.2 shows the average effectiveness ratio and an average across all COCOMs, with each COCOM receiving equal weight (20 percent).

It is clear from this table that the added effectiveness of multipoint refueling in deployment is highly dependent on the number of fighter aircraft that can be dragged by a single tanker.

Table 6.2
Effectiveness Ratios for Deployment to Various Theaters, Multipoint/Single-Point

COCOM	8 Fighters	12 Fighters
USAFRICOM	1.16	1.41
USCENTCOM	1.15	1.40
USEUCOM	1.16	1.43
USPACOM	1.15	1.40
USSOUTHCOM	1.17	1.47
Equal weight average	1.16	1.42

---

[3] These are the same proportions used in RAND's KC-135 recapitalization AoA (Kennedy et al., 2006). The proportions were based on the CAF force structure for 2025, as outlined in Pinter, 2004.

Although we present the increased effectiveness for the 12-fighter case, discussions with Air Mobility Command (AMC) personnel indicate that the maximum number of fighters per tanker, due to safety concerns may be either six or eight. If the number is six, then no benefit can be achieved on the deployment mission by adding multipoint capability to the fleet. In the cost-effectiveness analysis later in this section, we consider the potential benefit of using multipoint refueling on both the six- and eight-fighter cases. Again, referring to the analysis conducted in the KC-135 recapitalization AoA (Kennedy et al., 2006), not all fighters are multipoint-capable. In all cases, we assume that 15 percent of the receivers are probe-equipped and therefore multipoint-capable. In this case, the increase in effectiveness is zero for the six-fighter case and 2.4 percent for the eight-fighter case (0.15 × 16 percent = 2.4 percent). As discussed earlier, we do not comment in this analysis on the technical or operational feasibility of multipoint refueling to receptacle-equipped aircraft (i.e., boom pods). But this analysis shows the potential increase in effectiveness from the capability to conduct multipoint refueling for 100 percent of the fighter fleet.

Employment Mission

For the multipoint analysis, we considered the most demanding tanker orbit, orbit A, described in Chapter Three. Orbit A supported aircraft conducting air superiority, airborne electronic attack, and close controlled strike.

We assumed a two-minute hookup time for all refuelings (both single-point and multipoint) before fuel begins to flow. Once fuel begins to flow, we assumed that the flow rate on all fighters except the F-35 for single-point missions was 420 gallons per minute. For the F-35 in single-point operations, we assumed 320 gallons per minute.[4] In the multipoint case, we assumed that the flow rate for all fighter aircraft was 300 gallons per minute during refueling operations.

Given the number of fighters supported by this refueling orbit location, the hookup times, and the maximum fuel flow rates, it is not possible to meet this refueling demand with one tanker orbit. That is, even with perfect scheduling of fighters, it is not possible for one tanker at this location to meet the demand. By *perfect scheduling*, we mean that even if the fighters were "lined up" and waiting for their turn on the boom or basket, the first fighter refueled would reach a critical level of fuel before the last fighter had been refueled. This is not the way refueling operations are conducted. With the exception of packages of fighters, tight scheduling of refueling is not done because the goal is to reduce or eliminate wait time for fighter aircraft. Figure 6.6 shows the amount of time that the refueling point is occupied for different numbers of tankers simultaneously supporting this operation. We refer to this as *boom-occupied time*. Of course, boom-occupied time over 100 percent is impossible. As is expected, the boom-occupied time is inversely proportional to the number of tankers simultaneously supporting a tanker location (e.g., two single-point tankers per orbit location have their booms occupied 72 percent of the time, while four have their booms occupied 36 percent of the time).

Next, we determined the number of tankers required for a given number of simultaneous tanker orbits, as shown in Figure 6.7. The total number of tankers required was based on standard planning factors for times and aircraft speeds. We assumed that the tankers would transit the distance between the base and tanker orbit at 30,000 ft and mach 0.825. We used

---

[4] Meetings and conversations with staff at Lockheed Martin Aeronautics Company, December 18, 2009.

Figure 6.6
Boom-Occupied Time for a Given Number of Tankers Simultaneously Supporting a Refueling Orbit

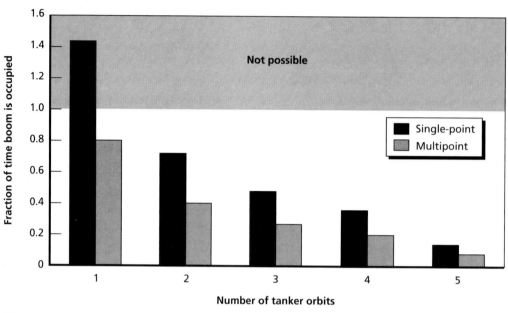

RAND *TR901-6.6*

Figure 6.7
Number of Tanker Aircraft Required Versus Number of Simultaneous Tanker Orbits

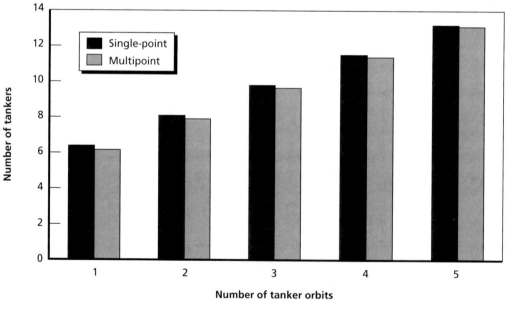

RAND *TR901-6.7*

the KC-10 performance manual to compute fuel used and times for all phases of the operation (climb, cruise, orbit). We assumed 15 minutes for start engines, taxi, and takeoff and 15 minutes for approach and land. We also assumed a ground time of four hours to account for turning the aircraft, not including the time required to refuel the KC-10. We assumed ten minutes to hook up the ground refueling apparatus and then computed the time required to refuel, assuming that fuel would flow at 4,000 lb per minute. This allowed us to compute the total sortie time from takeoff until the tanker was ready to take off for another sortie. With this information, we computed the number of aircraft required to support the operation. Figure 6.7 shows that operating this refueling orbit with two simultaneous tanker orbits requires about eight tankers devoted to the operation. Four simultaneous tanker orbits requires nearly 12 tankers to support the operation.

As shown in Figure 6.7, the number of tankers required to support a given operation depends on the number of simultaneous tanker orbits flown. The number of simultaneous tanker orbits required, in turn, depends on the amount of time one can reasonably assume the boom is occupied. Figure 6.6 presented the expected boom-occupied time under perfect conditions for the tanker—that is, perfect scheduling of receiver aircraft on the boom. This is not the way refueling operations are conducted. First, this scenario would require fighters to be lined up and waiting for their turn on the boom. This means that more fighters would be required to support a given level of fighter operations (e.g., a given number of simultaneous CAPs). Further, it would require perfect scheduling of fighters, which is not possible under operational conditions. As a result of the practical realities of operations, the fraction of boom-occupied time is dependent on the type of operation conducted, distances that receivers fly after refueling, time on station, number of receivers supported by each tanker, and other factors.

The reason that the boom-occupied time is of such importance in this analysis is due to the integer effects of the number of tankers required. From an examination of Figures 6.6 and 6.7, one can see that different levels of acceptable boom-occupied times greatly affect the equivalency ratio of multipoint to single-point effectiveness. Figure 6.8 presents this ratio for a range of maximum boom-occupied times. The ratio represents the number of tankers required

Figure 6.8
Effectiveness Ratio and Fraction of Boom-Occupied Time

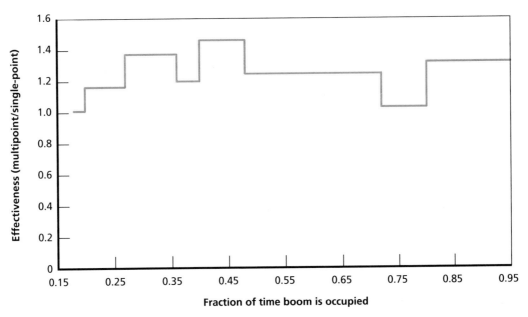

to support the operation; the steps result from the integer effects as the maximum acceptable boom-occupied time changes. Figure 6.8 shows that there are significant differences in the effectiveness ratio (multipoint/single-point), depending on the acceptable level of maximum boom-occupied time. The range goes from about 1.0 (representing no benefit from multipoint refueling) to more than 1.4, where a multipoint tanker is over 40 percent more effective than a single-point tanker. Complicating the issue is that there are large differences in the effectiveness ratio throughout the range of boom-occupied times analyzed. Higher boom-occupied times indicate more efficient scheduling of tankers, since there is less time between refuelings.

So, the question becomes, What level of boom-occupied time is reasonable to assume for real operations? For more insight, we analyzed the first day of OIF. We computed the amount of time that each tanker was refueling a fighter aircraft and divided it by the amount of time the tanker was on-station. Although we do not present those data here, we determined the number of sorties for each boom-occupied percentage. We then used this distribution to estimate the effectiveness of multipoint versus single-point refueling. Using a weighted average, we computed an effectiveness ratio of 1.25.

Similar to the deployment case, not all fighter sorties are probe-equipped and thus multipoint-capable. Again, consistent with RAND's KC-135 recapitalization AoA (Kennedy et al., 2006), we assumed that 35 percent of the employment fighter sorties would be capable of multipoint refueling. The employment mission uses 35 percent to account for additional allied and Navy aircraft that self-deploy; in the deployment mission, we used the assumption that 15 percent were multipoint-capable.

The effectiveness ratios we calculated for both the deployment and employment missions align with an improvement range of 17–50 percent, as found in Killingsworth's (1996) historical assessment of multipoint AR studies. The studies he reviewed used specific employment scenarios ranging from representative mission sets to an entire theater operation. The parametric approach of our analysis offers a framework for understanding the relationship of the particular scenarios to the effectiveness of multipoint refueling in the previous studies.

Other Operational Benefits

A potential operational benefit that is not reflected in the cost-effectiveness calculation is that multipoint refueling could enable greater fighter range after refueling. The maximum range of a fighter package is determined by the fighter with the least fuel. Multipoint refueling permits the fighters to refuel faster and reduces the amount of time that the first fighter to receive fuel must wait for the last fighter to receive fuel. We evaluated a case of an F-35A package with an air-to-ground load consisting of eight small-diameter bombs, two AIM-120s, and two AIM-9Xs, shown in Figure 6.9. The package refueled on both ingress and egress at 300 nm from base. It spends 20 minutes in the target area, and all weapons are expended. We also assumed a 500 nm divert requirement and 10-percent fuel reserve. Further, we also assumed that each fighter was refueled only once on both ingress and egress, with no topping off.

We used the same refueling assumptions in this part of the analysis that were used earlier. A two-minute hookup time was assumed. Refueling flow rates were 320 gallons per minute for the single-point refueling case and 300 gallons per minute for the multipoint case. Figure 6.10 shows the range of the package as a function of the number of aircraft for both the single-point and multipoint cases.

Figure 6.9
Tanker Orbit, Fighter Geometry

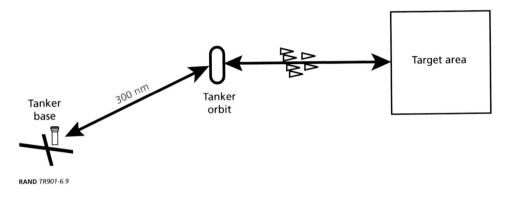

Figure 6.10
F-35 Fighter Package Strike Range for Multipoint and Single-Point Refueling

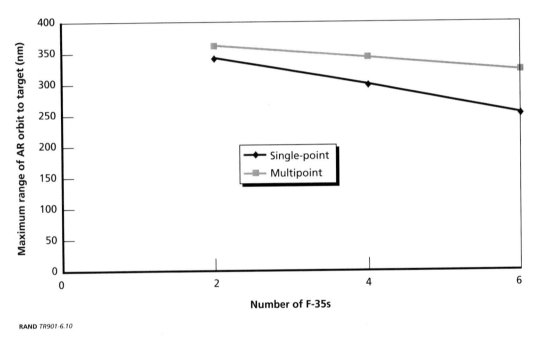

The figure shows that the potential fighter range increases with multipoint refueling as package size increases. A four-ship of F-35s can penetrate about 40 nm deeper using multipoint rather than single-point refueling. Another operational benefit is greater interoperability with allied aircraft, a higher percentage of which are probe-equipped than U.S. aircraft. This would also ease the wartime planning and allocation processes.

Additional Multipoint Refueling Capability Cost-Benefit Summary

In this section, we summarize the costs and calculated benefits for upgrading the KC-10 with additional multipoint refueling capability. All costs and benefits are in terms of NPV. NPV calculations take into account the time value of money and should be thought of as incorporating associated costs and benefits over the remaining life of the aircraft.[5] Using the methodology discussed earlier, we determined the procurement cost of adding multipoint refueling capability to each tanker as $4.2 million.

For the deployment mission, we considered cases of six, eight, and 12 receivers per tanker and assumed that 15 percent of the receivers would be multipoint-capable. For the employment mission, we considered boom-occupied time and assumed that 35 percent of receivers would be multipoint-capable. The resulting increases in wartime effectiveness are shown in Table 6.3. The results of our NPV benefit analysis are shown in Table 6.4. Comparing the benefits in Table 6.4 with the AUPC of $4.2 million, we see that the benefit exceeds the costs in all but the deployment-heavy mission mix constrained to a maximum of six receivers (see Figure 6.5).

Table 6.5 shows the cost-benefit summary for additional multipoint capability with the number of receivers for the deployment mission constrained to eight in consideration of the debate surrounding the maximum number of receivers allowed in a tanker formation.

Table 6.3
Effectiveness Changes Resulting from Additional Multipoint Refueling

Number of Receivers	KC-10 Wartime Effectiveness Δ (%)	
	Employment	Deployment
6	8.8	0
8	8.8	2.4
12	8.8	6.7

Table 6.4
Benefit of Additional Multipoint Refueling

Number of Receivers	Benefit (FY 2009 $ millions)	
	Employment-Heavy	Deployment-Heavy
6	13.6	3.3
8	15.5	7.6
12	18.9	15.2

Table 6.5
Additional Multipoint Refueling Capability Cost-Benefit Summary

AUPC (FY 2009 $ millions)	Mission Mix	NPV Benefit (FY 2009 $ millions)
4.2	Employment-heavy	15.5
4.2	Deployment-heavy	7.6
4.2	Average benefit	11.6

---

[5] NPV is the appropriate way to judge modernization investments based on the true resource cost. OMB directs this kind of discounted analysis. We used the current (December 2009) OMB-directed real long-term discount rate of 2.7 percent in our analysis, which represents the return on investment (OMB, 2009).

CHAPTER SEVEN

Defensive Systems

Background

The KC-10, like the Air Force's other large tanker, the KC-135, does not currently have defensive systems. Defensive systems would decrease the vulnerability of the KC-10 by lowering the probability that threats are able to hit the aircraft during operations, increase crew situational awareness of potential and incoming threats, and, should threats hit the aircraft, increase the probability of aircraft survival.

The threats faced by tanker aircraft are dependent on whether the aircraft is at low altitudes (e.g., during takeoff and landing) or at high altitudes, as during AR or en route cruise. During low-altitude flight, the primary threat to all mobility aircraft, including tankers, is MANPADS. MANPADS generally use IR guidance systems to home in on their targets. Other threats at low altitudes include small arms and unguided rockets and missiles. At higher altitudes, the threats include larger surface-to-air missile (SAM) systems that are typically guided using radar or radio frequency (RF) guidance systems. Other threats to tankers at all altitudes stem from antiaircraft artillery and air interceptor aircraft, which could attack by employing a range of unguided IR or RF weaponry. Protection from air interceptors for tanker aircraft would not typically come from an onboard system but, rather, would be provided by friendly defensive fighter aircraft, like the F-15 and F-22.

Figure 7.1 shows the percentage of U.S. aircraft lost or damaged in combat since 1991 by threat type. Most of the aircraft that were hit had been targeted by antiaircraft artillery systems, followed by IR- and then RF-guided SAMs. Figure 7.2 shows the number and type of aircraft that were either damaged or lost by the threats depicted in Figure 7.1. Note that not one tanker aircraft has been either damaged or destroyed by enemy fire.

Tanker losses have been avoided through a combination of air superiority, basing tanker aircraft at relatively low-threat airfields, and conducting AR operations outside the weapon engagement zone of known threats.

Even though no tankers have been lost to enemy fire, they still often enter the MANPADS weapon engagement zone when taking off or returning to base (Figure 7.3). To avoid the MANPADS threat, tankers typically have been based far from refueling orbits. While conducting AR operations, tanker aircraft are usually above the engagement envelope of MANPADS and small-arms fire but are within the altitude envelope of many strategic SAM systems (Figure 7.4). To avoid these large land-based SAM systems, AR locations are established laterally outside the engagement envelope of known SAM sites.

Defensive systems would primarily affect the tanker's wartime missions of employment and, potentially, deployment. For employment, the effect of defensive systems on the mission

Figure 7.1
Percentage of Aircraft Damaged or Destroyed,
by Threat Type

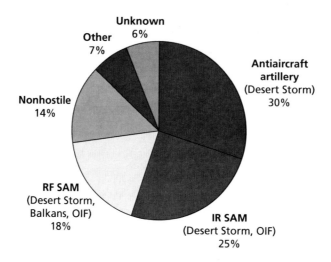

NOTE: N = 111. Two of three aircraft hits in OIF were from
PATRIOT batteries.

RAND *TR901-7.1*

Figure 7.2
Number of Fixed-Wing Aircraft Lost or Damaged in Combat, by Type, 1991–Present

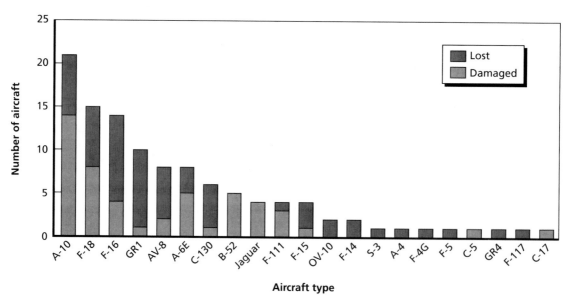

NOTE: N = 111.

RAND *TR901-7.2*

will differ based on the air defense capabilities of the adversary. In a conflict with intact enemy air defenses and no allied air superiority, defensive system–equipped tankers may be able to fly closer to threats, allowing mission receiver aircraft to penetrate deeper into enemy territory. In a situation in which the United States enjoys allied air superiority and minimal enemy

Figure 7.3
Typical Dimension of MANPADS Engagement Envelopes During Takeoff and Landing

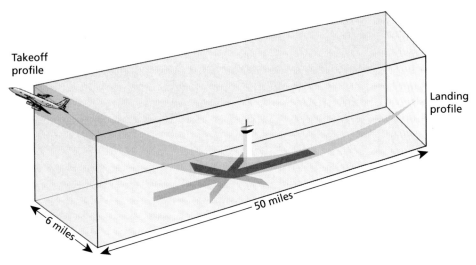

SOURCE: Northrop Grumman, undated. Used with permission.

RAND TR901-7.3

Figure 7.4
Typical AR Altitude Compared with SAM Engagement Zones

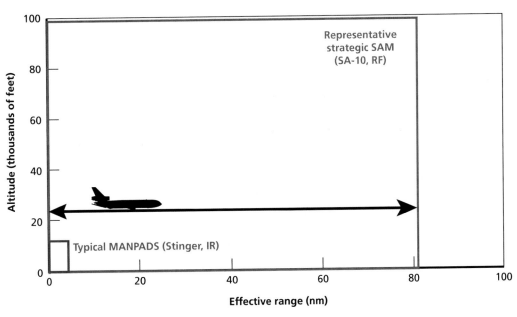

RAND TR901-7.4

air defenses, defensive systems may allow aircraft to be based closer to refueling locations, reducing the number of tankers required and increasing the fuel available for offload. For the deployment mission, a defensive system–equipped tanker could take on an expanded mission of delivering fighter-support equipment to the deployed location in addition to escorting and refueling the fighter aircraft. Along the same lines, tankers could be used in an expanded airlift role, operating at a similar risk level to airlift aircraft equipped with defensive systems.

There is a wide range of defensive systems currently in production and installed on other large aircraft. These systems can be thought of in terms of the function they provide. For this study, we examined adding to the KC-10 the types of defensive systems that are found on other large aircraft in the U.S. Air Force inventory, like the C-17, C-5, and C-130, including systems for missile protection, crew protection, and fuel tank protection. Missile protection systems fall into two broad categories: IR threat protection and RF threat protection, depending on the guidance systems of the missiles they are trying to protect against. Crew protection in large aircraft typically consists of adding armored panels to the interior of the aircraft to protect the crew against small-arms fire and shrapnel fragments. The last category we considered was fuel tank protection, which uses different methods to reduce the possibility of explosions in aircraft fuel tanks.

In our estimation, this is a reasonable set of components for AR and airlift missions conducted by a subsonic, nonstealthy aircraft like the KC-10.

Missile Protection

SAM systems have proliferated widely around the world, with a vast majority of countries operating MANPAD-based systems and a large number possessing the larger strategic SAM systems. Figures 7.5 and 7.6 show the worldwide proliferation of MANPADS and strategic-type SAM systems, respectively. Neither of these figures accounts for nonstate actors or groups that may have such systems.

Clearly, based on Figures 7.5 and 7.6, there is a threat from MANPADS or SAM systems in most places in which tanker aircraft are likely to operate. Figure 7.5 and 7.6 are meant to show the worldwide proliferation of these systems, not to imply that there would be a threat from host-nation forces where KC-10s may be based. Rather, the largest threat to large mobil-

Figure 7.5
Worldwide MANPADS Proliferation

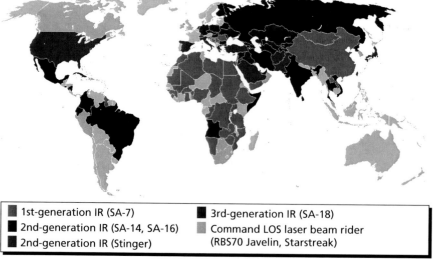

- 1st-generation IR (SA-7)
- 2nd-generation IR (SA-14, SA-16)
- 2nd-generation IR (Stinger)
- 3rd-generation IR (SA-18)
- Command LOS laser beam rider (RBS70 Javelin, Starstreak)

SOURCE: Data from *Jane's Land-Based Air Defence*.
NOTE: Many countries and groups possessing later-generation missiles also possess earlier-generation weapons.

RAND *TR901-7.5*

Figure 7.6
Worldwide Strategic SAM Proliferation

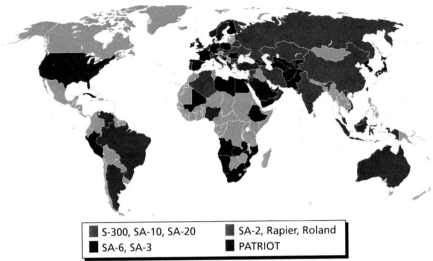

■ S-300, SA-10, SA-20	■ SA-2, Rapier, Roland
■ SA-6, SA-3	■ PATRIOT

SOURCE: Data from *Jane's Land-Based Air Defence*.
NOTE: Many countries and groups possessing later-generation missiles also possess earlier-generation weapons.

RAND *TR901-7.6*

ity aircraft, like tankers, will be from MANPADS operated by terrorists, insurgents, or very small enemy special forces units attacking the aircraft during takeoff and landing. These types of attacks are illustrated by the MANPADS hits on a DHL Airbus 300 in November 2003, a C-5 in December 2003, and a C-17 in January 2004 (Schroeder, 2007). Protection from these systems typically comes from defeating the guidance systems of the incoming missiles.

To protect against IR missile systems, like most MANPADS, aircraft are often equipped with a missile warning and countermeasure system. Missile warning systems detect incoming missiles by sensing their IR or ultraviolet energy signature. This information is often provided to the aircrew as well as to a countermeasure system.

Two countermeasure systems currently in use on military aircraft are pyrotechnic flares and laser-based systems. The flares are ejected from dispenser units and act as decoys, attracting incoming missiles with large amounts of IR energy in specifically designed spectrums and ejection patterns. Laser-based countermeasures, like the Large Aircraft Infrared Countermeasures System (LAIRCM), take the information from the missile warning system and use a laser to disrupt the incoming missile's guidance system (AFOTEC, 2007).

Protection from RF-guided missiles is similar, but detection is provided by specialized receivers that detect the radar energy from either land- or missile-based radars that are used to guide SAMs to incoming aircraft. RF countermeasure systems include metallic chaff and towed decoy systems. Chaff functions analogously to flares against IR threats. In the case of chaff, strips of radar reflective material are dispensed from the aircraft, act as large reflectors to guidance radar, and provide a much larger radar target than the aircraft itself. Also, like flares, the chaff is specifically designed to be effective against a large range of guidance radars. Towed decoys are trailed behind the aircraft they protect and emit signals to both jam and confuse the radar of the incoming missile system, increasing the chance that the missile will miss the aircraft.

We selected as our candidate missile protection suite a LAIRCM-based IR protection system with five detectors, three laser turrets, and a radar warning receiver. This system provides protection similar to that of other large U.S. Air Force aircraft, such as the C-17, C-5 and C-130, and British Royal Air Force tankers (Office of the Under Secretary of Defense [Comptroller], 2010). An RF countermeasures system was not included because of the difficulties associated with protecting a subsonic aircraft with a large radar cross-section that requires the area in the rear of the aircraft to be clear to conduct refueling operations.

Crew Protection

Crew protection is intended to shield the aircrew from both small-arms fire and shrapnel fragments. This protection comes from panels of a Kevlar-based material that can be installed around the inside flooring of the aircraft cockpit and other vulnerable areas. We included an armor system in our defensive system package that is comparable to the armor installed in other AMC aircraft.

Fuel Tank Protection

Fuel tank protection seeks to minimize the possibility of a fuel tank explosion should there be a fire or if a tank is hit by enemy threat systems. There are three main types of fuel tank protection systems: onboard inert gas–generating systems, liquid nitrogen systems, and internal fuel tank foam. For our analysis, we included an onboard inert gas–generating system. We eliminated the liquid nitrogen system because it is dependant on the local availability of liquid nitrogen. The internal foam was eliminated because it traps a significant percentage of fuel, which would increase the tanker's weight and reduce its offload by as much as 25,000 lb. Additionally, the internal foam increases the maintenance cost of the fuel tanks because each time the tanks are opened, a portion of the foam must be removed and replaced.

Upgrade Costs

The overall AUPC of the defensive system in our analysis was $19.2 million per TAI, broken down according to the percentages shown in Figure 7.7.

Table 7.1 shows two contractor estimates of the flyaway shipset cost for LAIRCM,[1] an estimate by the U.S. Department of Homeland Security (DHS) for equipping civilian airliners with this system (DHS, 2006, Section 5), and the budget cost of procuring and installing LAIRCM on the C-5 and C-17. The two contractor estimates are close, averaging $9.1 million. The budget costs for the C-5 and C-17 are in this range as well. The DHS cost is much lower, by half an order of magnitude. This cost is estimated based on equipping a very large number of aircraft with the system and benefits from the very large economies of scale expected to result from such a large program. We judge that the contractor cost estimates are reasonable in light of the C-5 and C-17 budget costs and our expectation that a very widespread commercial program will not occur. If one did, the budget cost for it should be used to revise our estimate. We used the average of the two contractor costs as our estimate in this study.

---

[1]   We show contractor estimates as "Contractor A," "Contractor B," and so on, for each modernization option. These designations do not necessarily refer to the same contractor in different tables.

Figure 7.7
Relative Proportions of Defensive System Component Costs

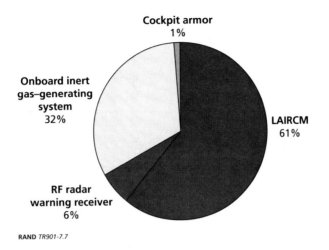

RAND *TR901-7.7*

Table 7.1
LAIRCM Upgrade Cost Estimate Sources

Source of Cost Estimate	Flyaway Cost per Shipset (FY 2009 $ millions)
Contractor A	9.57
Contractor B	8.69
DHS	1.83
C-5 (PB 2009)	7.19
C-17 (PB 2005–2009)	6.84

We applied a 35-percent flyaway-to-AUPC markup to that estimate, based on the C-5, C-17, and C-130 LAIRCM PB data.[2]

Table 7.2 shows two contractor estimates of the flyaway shipset cost for the radar warning receiver and the budget cost of procuring and installing it on the C-130E/H. The two contractor estimates differ by almost a factor of two, and both include more equipment than the ALR-69A, the system included in this study. The ALR-69A was installed as part of the C-130E/H radar warning receiver program, so we judged that the C-130E/H budget cost was the best analogy and used it for this study. We also applied the standard 15-percent flyaway-to-AUPC markup.

Table 7.3 shows two contractor estimates of the flyaway shipset cost for the onboard inert gas–generation system and the budget cost of procuring and installing it on the C-17. The two contractor estimates differ by more than an order of magnitude. We judged that the C-17

---

[2] The support factors for the C-5, C-17, and C-130 programs are 55 percent, 33 percent, and 30 percent, respectively. Data from all programs are confounded by various considerations. The C-5 P3A covers only the first 14 aircraft. The C-17 case is complicated by (1) the change from the LAIRCM "lite" to the LAIRCM "full" configuration, which has more turrets, and (2) upgrading from the small laser to the Guardian laser. The C-130 case is complicated by the transfer of some equipment that was removed from C-17s. Our judgment is that 35 percent is a representative value.

Table 7.2
Radar Warning Receiver Cost Estimate Sources

Source of Cost Estimate	Flyaway Cost per Shipset (FY 2009 $ millions)
Contractor A	1.71
Contractor B	3.29
C-130E/H (PB 1997–2009)	0.96

Table 7.3
Onboard Inert Gas–Generating System Cost Estimate Sources

Source of Cost Estimate	Flyaway Cost per Shipset (FY 2009 $ millions)
Contractor A	0.78
Contractor B	15.27
C-17 (PB 2009)	4.85

budget cost was the best analogy in this case and used it for this study. We also applied the standard 15-percent flyaway-to-AUPC markup.

Table 7.4 shows the budget cost of procuring and installing cockpit armor on the C-130 and the C-17, as well as the cost reported for this modification on the C-5 as stated in a contract award notice.[3] The C-130 budget cost and the C-5 contract award notice cost are almost the same. The C-17 budget cost is almost an order of magnitude higher; it comes from the FY 2009 PB P-40 sheet, which has no details on cost structure, and funding does not begin until FY 2011. We judged that the C-130/C-5 budget costs were the best analogies in this case and used the average for this study. We also applied the standard 15-percent flyaway-to-AUPC markup.

Changes to Peacetime Operations Costs

If a LAIRCM system is installed, there will be additional aerodynamic drag that will increase peacetime fuel usage. The laser turret and sensors protruding into the airstream cause the drag,

Table 7.4
Cockpit Armor Cost Estimate Sources

Source of Cost Estimate	Flyaway Cost per Shipset (FY 2009 $ millions)
C-130 (PB 2003–2004)	0.18
C-5 contract award notice	0.20
C-17 (PB 2009)	0.92

---

[3] The C-5 estimate is based on the January 2008 contract awarded to Foster Miller-Last Armor (see Office of the Assistant Secretary of Defense [Public Affairs], 2008).

which would probably result in an increase in fuel usage of between 1 and 2 percent. Based on a 2-percent fuel usage increase, this amounts to $2.2 million per TAI per year. Two percent was chosen as a conservative number based on C-17 performance data, with and without LAIRCM, and DHS testing of LAIRCM systems on commercial cargo-carrying aircraft.

We used the 1996–2006 flying hour and fuel use averages for the KC-10 fleet based on the Air Force Total Ownership Cost database to calculate the value of the 2-percent fuel burn increase. For our analysis, we project that KC-10s will fly 953 hours per year and burn 2,817 gallons of jet fuel per flying hour. Thus, each KC-10 burns 2.68 million gallons per year; a 2-percent increase translates to 53,700 gallons. We project the price of jet fuel to be $2 per gallon in dollars of FY 2009 purchasing power, so the additional annual cost of a 2-percent increase in fuel use per flying hour is $107,000.

In this study, we assumed that the KC-10s would be operated for an average of 30 years after the modernizations are installed. We must therefore include the value of the all the fuel cost increases over this period. For this, we used the present value of the total future stream of cost increases; that is, the one-time outlay, occurring at the same time that the upgrade occurs, that would have the same value as that of all of the future outlays. OMB requires that the U.S. government use the present-value method for valuing costs that occur over time and has specified that a 2.7-percent real interest rate be used for periods of 30 years (OMB, 2009). At a real interest rate of 2.7 percent, a cost of $107,000 incurred every year for 30 years has a present value of $2.2 million. This is less than the total amount of the real outlays, which is 30 × $0.107 million = $3.2 million. Outlays that occur later have lower costs than those that occur earlier because of the 2.7-percent real return to saving that is embodied in the present-value approach.

The total cost of the defensive system modernization option is therefore $21.4 million, the sum of $19.2 million, the acquisition cost, and the $2.2 million operations cost.

Valuing Wartime Effectiveness

There are potential changes in the number of KC-10s required for wartime based on the addition of defensive systems to the aircraft. As described earlier, defensive systems may allow KC-10s to be based closer to their wartime refueling locations and may allow them to conduct AR operations in locations closer to known threats.

The value of moving bases closer to their refueling locations depends on both the mission mix and change in distance. We assumed that the KC-10s would be based 200 nm closer to their AR locations. The rationale for the 200 nm stems from deployed tanker locations in OIF. In 2003, if KC-10s were based at locations where coalition aircraft with defensive systems operated (instead of their actual locations), the average reduction in distance to the AR areas would have been 200 nm. It should be noted that tanker basing is not just a function of threat and threat-mitigation options. Rather, several other factors also contribute to aircraft and tanker basing decisions. These factors include, but are not limited to, fuel availability, ramp space available to park the aircraft, ability of the location to support the personnel required to operate the aircraft, and the relative ranges of different aircraft in the theater. Nonetheless, the addition of defensive systems would provide the KC-10 with the protection that it currently lacks and would equip it with systems that are required on other Air Force aircraft for access to specific higher-threat locations.

Under the employment-heavy mission mix, we found that for defensive systems to be cost-effective, the tankers would have to be based 255 nm closer to their AR locations. Under the deployment-heavy mission mix, the defensive systems for the KC-10 are not cost-effective, even if the tanker bases are located directly under the refueling locations. This outcome results from the fact that there is not enough of the employment mission to offset the cost of the defensive systems.

In our judgment, even with defensive systems, tankers and KC-10s would not operate in AR orbits closer to the threat. As a result, there is no change in the number of tankers required in this scenario. Were the tankers allowed to operate closer to known threats, more tankers may be required because the tankers would be flying farther from their bases and supporting other combat aircraft, which would also be flying farther, as described in Chapter Five. While moving the AR locations of the tankers farther forward may provide operational benefits not currently in the fleet, it increases the number of tankers required and thus would not pay a cost benefit.

Other Operational Benefits

We identified two other operational benefits from operating tankers equipped with defensive systems.[4] Tankers flying over high terrain would be at reduced risk if equipped with defensive systems. Defensive systems would also allow KC-10s to operate more as airlifters and in a dual tanker-airlift role.

Reduced Risk Refueling over High Terrain

AR occurs at altitudes that are compatible with both tanker and receiver aircraft performance. For a given receiver and gross weight, there is an optimal pressure altitude for AR. Ground-based threats are not as dependant on pressure altitude and typically are able to reach a given altitude based on the elevation of the terrain from which they were fired. So, a shoulder-fired missile launched from a 10,000 ft mountaintop may be able to strike aircraft operating at altitudes over 20,000 ft above mean sea level—for example, when conducting AR operations over mountainous countries, like Afghanistan, where a significant amount of the terrain is above 10,000 ft.

Increased Operation as a Dual-Role Tanker or Airlifter

The KC-10 has significant cargo and passenger capability that allows it to operate in a pure airlift role and in a dual tanker-airlift role. In the pure airlift role, the KC-10 is only moving cargo, personnel, or both from location to location. In the dual tanker-airlift role, the KC-10 acts as an escort tanker for deploying fighter-sized aircraft while simultaneously carrying the fighters' support equipment and personnel as cargo. An FY 2006–2007 snapshot of KC-10 flying hours shows minimal use of the dual-role capability—1 to 2 percent of flying hours and between 10 and 15 percent of flying hours in an airlift role (Air Mobility Command, 2008).

Previous studies (for example, Kennedy et al., 2006, and DoD and JCS, 2005) have shown that the most efficient concept of operations in terms of the number of tankers required to deploy fighter aircraft is the round-robin concept described in Chapters Three and Six.

---

[4] As one of our reviewers noted, another benefit of adding this suite of defensive systems is as a hedge against the increased capability of adversary weapon systems.

However, in the round-robin concept of operations, the tankers do not actually land at the fighters' deployed basing location. Rather, they escort the fighter aircraft to a point where the fighters have the range to make it to the deployed location on their own. The support equipment and personnel for the fighters is delivered to the deployed fighter base location by airlift aircraft, which are equipped with defensive systems because of the potential threat at the deployed location. Tanker aircraft equipped with defensive systems would also be able to land at the deployed location with a risk similar to that of airlifters equipped with defensive systems, which would allow tankers to be used as airlifters or in a dual role to deploy fighter aircraft. We present a comparison of the number of tankers and airlifters required using three concepts of delivering fighters and their support equipment and personnel to theater.

Each of the scenarios involves the deployment of a package of 24 F-22s. We assume that these 24 F-22s require 258 support personnel and 108 (4,000 lb) pallets of support equipment, based on information provided by the AMC Analysis, Assessments, and Lessons Learned Directorate (AMC/A9).

We compared three concepts of operations (CONOPS) for deploying the 24 F-22s and their support equipment: baseline, pure airlift KC-10s, and dual-role KC-10s. In the baseline case, the KC-10s are used strictly as refueling aircraft to deploy the fighters using the round-robin approach described earlier (see Figure 7.8). C-17s deploy the support equipment and personnel. Figure 7.9 shows the number of KC-10s and C-17s required over a range of deployment distances for the baseline CONOPS. For a given distance, the number of C-17s required to move the personnel and cargo is roughly double the number of KC-10s required to deploy the fighter aircraft. We use these numbers as a benchmark to compare the other two scenarios.

For the case of the KC-10 in an airlift role, some KC-10s are still used strictly as refueling aircraft to deploy the fighters using the round-robin CONOPS. However, instead of C-17s deploying the personnel and equipment, additional KC-10s operating in a pure airlift role

Figure 7.8
Baseline Deployment CONOPS

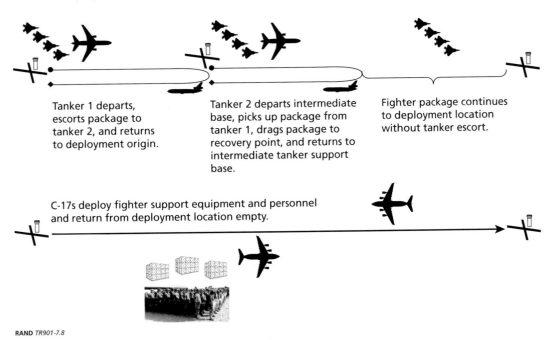

Tanker 1 departs, escorts package to tanker 2, and returns to deployment origin.

Tanker 2 departs intermediate base, picks up package from tanker 1, drags package to recovery point, and returns to intermediate tanker support base.

Fighter package continues to deployment location without tanker escort.

C-17s deploy fighter support equipment and personnel and return from deployment location empty.

Figure 7.9
Number of KC-10s and C-17s Required to Deploy 24 F-22s in Baseline CONOPS

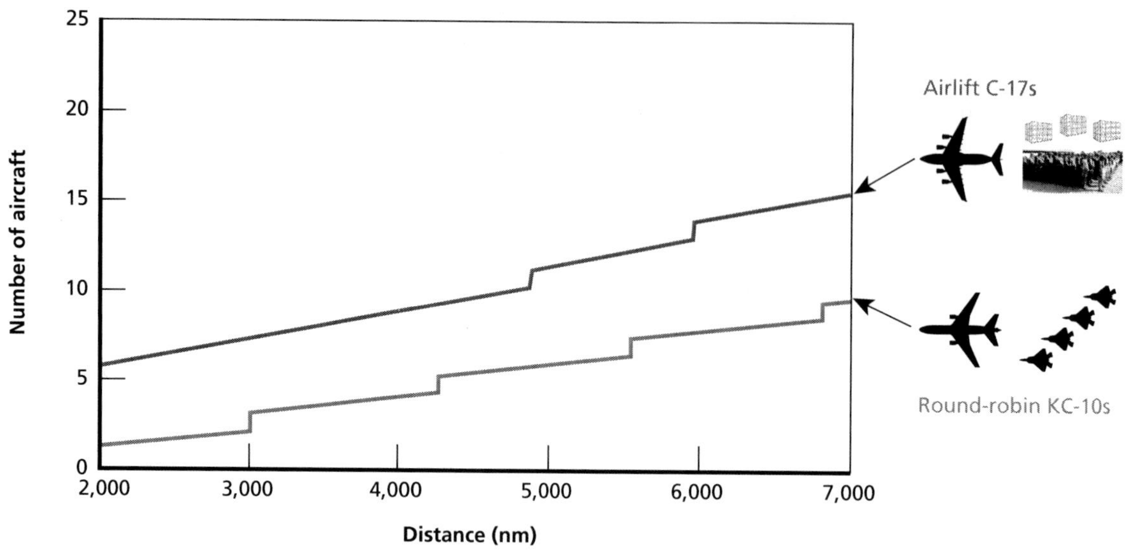

deliver the support equipment and personnel. For the purpose of illustration, we assume that all the support equipment would be transportable by KC-10s. In actuality, some amount of fighter support equipment is oversized and not transportable in a KC-10. For a deployment of 24 F-22s we estimate that at least two C-17–equivalent loads would need to be transported by some aircraft capable of handling this outsized cargo. We arrived at two C-17 loads by excluding each individual piece of cargo that was larger than two standard 463L pallets—essentially, the limit to fit through the KC-10 cargo door.

Figure 7.10 adds to the previous figure the number of KC-10s that would be required to airlift the support personnel and equipment. Note that, to complete the deployment of the entire package, the total number of aircraft required would be the *either* the sum of the round-robin KC-10s and airlift C-17s or the sum of the round-robin KC-10s and pure airlift KC-10s. Over all distances, the number of KC-10s required to move the support package is less than the number of C-17s required because of the KC-10's greater bulk cargo capacity and range compared with that of the C-17 for items that are commonly transportable by both aircraft. If KC-10s are required in theater for the employment mission, using their inherent airlift capacity on the way into theater improves KC-10 effectiveness.

For the dual-role case, KC-10s deploy the fighters directly to the deployment location while also carrying support equipment and personnel (see Figure 7.11). The amount of support equipment and personnel aboard each KC-10 is proportional to the number of fighters in trail. The dual-role KC-10s themselves are air refueled by additional KC-10s as needed, based on distance.

Figure 7.12 shows the total number of KC-10s required to deploy the F-22 package in the dual-role CONOPS. For comparison, the figure includes the number of round-robin KC-10s and airlift C-17s from the baseline case. The top line in Figure 7.12 is the sum of the dual-role and force-extension KC-10s required. In the figure, the total number of aircraft needed for the deployment would be the *either* the sum of the round-robin KC-10s and airlift C-17s or the total number of KC-10s (the top line in the figure). Using the KC-10 in the

Figure 7.10
Number of KC-10s and C-17s Required to Deploy 24 F-22s in Pure Airlift CONOPS

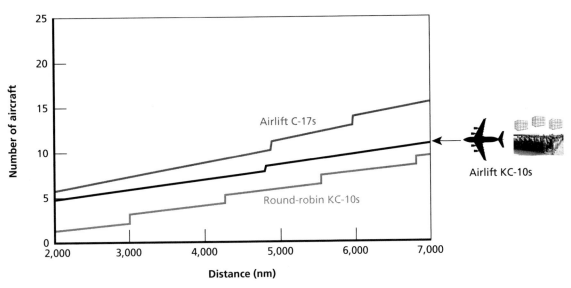

Figure 7.11
Dual-Role Deployment CONOPS

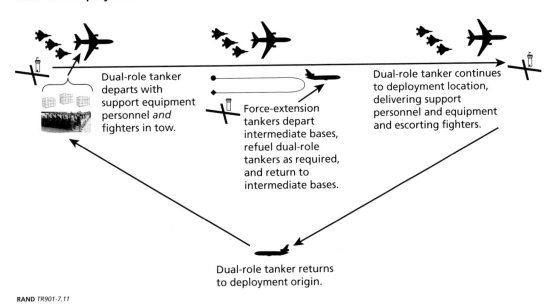

dual-role CONOPS requires an additional number of KC-10s over the baseline case, roughly equivalent to the number of C-17s that would be needed for the airlift of the support equipment and personnel. In the figure, this is represented by the two arrows. The first arrow shows the added number of KC-10s required over the baseline case, and the second shows the number of C-17s needed to airlift the support package.

Comparing the dual-role and pure airlift CONOPS shows that the pure airlift approach requires fewer additional KC-10s. This is in agreement with a previous study examining in the use of dual-role KC-10s (Hunsuck, 1986). To value the use of the KC-10 in these deployment

Figure 7.12
Number of KC-10s and C-17s Required to Deploy 24 F-22s in Dual-Role CONOPS

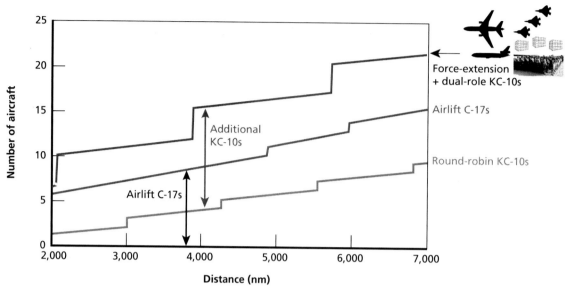

roles, we can compare the number of KC-10s with the number of C-17s that could be freed to do other missions. Note that we do not compare the procurement cost of the two aircraft because we assume that both aircraft are already purchased. However, should future tankers or airlifters need procuring, this should factor into the decision. In valuing the KC-10's contribution in this manner, we assume that the C-17s would be free to conduct other required missions for which the C-17 is the most cost-effective (for example, moving a large number of wheeled vehicles over a long distance).

Based on analyses conducted in RAND's KC-135 recapitalization AoA (Kennedy et al., 2006), the cost ratio of the KC-10 to the C-17 is 0.6 (based on NPV). If the number of C-17s replaced divided by the number of additional KC-10s required is greater than 0.6, the KC-10 CONOPS would be more cost-effective. Figure 7.13 shows that both the dual-role and pure airlift CONOPS are more cost-effective than using C-17s for the airlift of the support package in a fighter deployment. Even if the cost ratio were 1.0, using the KC-10 in a pure airlift role would be more cost-effective than using C-17s for the cargo they can both carry.

Defensive Systems Cost-Benefit Summary

In this section, we summarize the costs and calculated benefits for upgrading the KC-10 with the defensive system suite discussed in this chapter. All costs and benefits are in terms of NPV. NPV calculations take into account the time value of money and should be thought of as incorporating associated costs and benefits over the remaining life of the aircraft.[5] Table 7.5 presents the summary of costs for adding defensive systems to the KC-10.

---

[5]   NPV is the appropriate way to judge modernization investments based on the true resource cost. OMB directs this kind of discounted analysis. We used the current (December 2009) OMB-directed real long-term discount rate of 2.7 percent in our analysis, which represents the return on investment (OMB, 2009).

Figure 7.13
Ratio of Additional KC-10s to C-17s Needed for Deployment

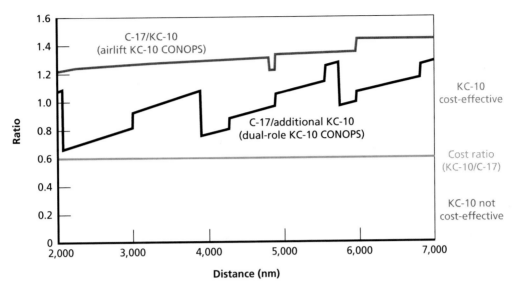

Table 7.5
Costs of Defensive Systems

System/Category	AUPC (FY 2009 $ millions)
LAIRCM	12.3
Radar warning receiver	1.1
Onboard inert gas–generating system	5.6
Cockpit armor	0.2
Total	19.2

The baseline assumptions used in the NPV estimates include a 200 nm reduction in the distance required for the tanker to travel during the employment mission, making it 10.6 percent more cost-effective in that mission.

Table 7.6 shows the resulting cost-benefit summary for the defensive system upgrade. The table accounts for the tanker savings from the employment mission only and does not incorporate any savings from using tankers to offset airlift in the deployment role or any benefit from the reduced risk when operating over high terrain.

The calculus of the cost-effectiveness of adding defensive systems would change if the cost of the installation and equipment were different from our estimate or if the tankers were based more or less than 200 nm away. Figure 7.14 shows the NPV parametrically given different costs of the defensive system suite and different distances for closer tanker basing in the employment-heavy mission mix. In Figure 7.14, the NPV on the vertical axis represents the benefit less the total cost, so positive values result in more benefit than cost and negative values cost more than the benefit provided. The condition chosen in this study ($19.2 million AUPC and 200 nm closer) is shown in the figure by the arrows and circle.

Table 7.6
Defensive Systems Cost-Benefit Summary

Type	Cost (FY 2009 $ millions)	Mission Mix	NPV Benefit (FY 2009 $ millions)
AUPC	19.2	—	—
Additional operating cost	2.2	Employment-heavy	16.4
		Deployment-heavy	4.0
Total	21.4	Average	10.2

Figure 7.14
NPV of Defensive Systems with Varying Equipage Costs and Basing Location Improvement

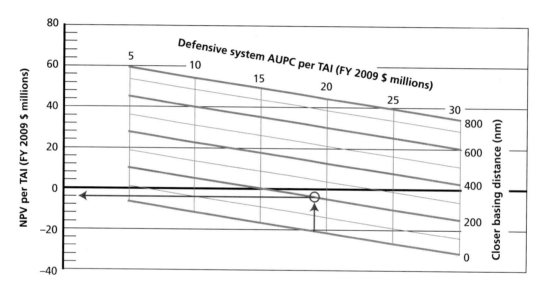

RAND TR901-7.14

CHAPTER EIGHT

Night Vision Imaging System–Compatible Lighting

Background

KC-10 lighting is not currently compatible with NVG operations. Incorporating compatible lighting would allow tanker crews to operate the aircraft in all mission phases while wearing NVG. In addition, it would allow receiver aircraft pilots to wear NVG while conducting refueling operations. It follows, then, that the cost-effectiveness of modifying the lighting of the KC-10 depends on the value to tanker and receiver crews operating with NVG. There are a few hypotheses regarding the value of NVIS-compatible lighting on tanker aircraft. One thought is that incorporating compatible lighting on the KC-10 exterior could allow receiver pilots to refuel without removing their NVG, potentially reducing the time required for refueling operations. Another hypothesis is that adding NVIS-compatible lighting to the cockpit and cargo area could allow basing and cargo operations at covertly lit or blacked-out airfields. This would allow the KC-10 to be unseen without night-vision devices while taking off, landing, and loading or unloading passengers and cargo while on the ground. Another suggested benefit of NVIS-compatible lighting is that it would increase safety at busy airports where there are a variety of aircraft operating in a mixed-lighting environment.[1] The last hypothesis is that night-vision devices would facilitate special operations forces AR. After presenting the costs to modify the KC-10 lighting, we explore each of these hypotheses.

Upgrade Costs

Our estimate to upgrade the lighting on the KC-10 to be NVIS-compatible is $3.6 million per TAI. This value includes upgrading the exterior, the air refueling operator (ARO) station, the cockpit, and the cargo area.

Table 8.1 shows two contractor estimates of the flyaway shipset cost for the exterior NVIS modification and the budget cost of procuring and installing it on the C-17.[2] The three estimates differ substantially. We judged that the average of the contractor estimates was the best estimate in this case and used it for this study. We also applied the standard 15-percent flyaway-to-AUPC markup.

---

[1] That is, where some aircraft are operating blacked out with no visible external lights and their pilots are using night-vision devices, while others may be operating with external lights out, but their pilots are not using night-vision devices.

[2] We show contractor estimates as "Contractor A," "Contractor B," and so on, for each modernization option. These designations do not necessarily refer to the same contractor in different tables.

Table 8.1
Exterior NVIS-Compatible Lighting Cost Estimate Sources

Source of Cost Estimate	Flyaway Cost per Shipset (FY 2009 $ millions)
Contractor A	0.52
Contractor B	2.66
C-17 (PB 2009)	0.37

Table 8.2 shows two contractor estimates of the flyaway shipset cost for the ARO station NVIS modification. They differ by an order of magnitude. With no budget data from actual programs to use as analogies, we used the average of the two contractor estimates for this study. We also applied the standard 15-percent flyaway-to-AUPC markup.

Table 8.3 shows two contractor estimates of the flyaway shipset cost for the cockpit NVIS modification. These estimates are similar, and we used their average for this study. We also applied the standard 15-percent flyaway-to-AUPC markup.

Table 8.4 shows two contractor estimates of the flyaway shipset cost for the cargo-area NVIS modification and the budget cost of procuring and installing it on the C-17. We judged that the average of the contractor estimates was the best estimate in this case and used it for this study. We also applied the standard 15-percent flyaway-to-AUPC markup.

Table 8.2
ARO Station NVIS-Compatible Lighting Cost Estimate Sources

Source of Cost Estimate	Flyaway Cost per Shipset (FY 2009 $ millions)
Contractor A	0.10
Contractor B	1.21

Table 8.3
Cockpit NVIS-Compatible Lighting Cost Estimate Sources

Source of Cost Estimate	Flyaway Cost per Shipset (FY 2009 $ millions)
Contractor A	0.82
Contractor B	0.97

Table 8.4
Cargo-Area NVIS-Compatible Lighting Cost Estimate Sources

Source of Cost Estimate	Flyaway Cost per Shipset (FY 2009 $ millions)
Contractor A	0.024
Contractor B	0.025
C-17 (PB 2009)	0.737

Changes to Peacetime Operations Costs

We did not include any changes to peacetime operating costs for NVIS-compatible lighting. It is conceivable that newer light-emitting diode–based systems could have a lower sustainment cost because these types of bulbs tend to last significantly longer than their incandescent predecessors.

Valuing Wartime Effectiveness

In this section, we examine the potential reduction in the number of tankers with the faster cycling of receivers through the use of NVIS-compatible lighting. All Air Combat Command pilots who employ NVG must discontinue using the devices when they conduct AR.[3] Some amount of time is required for their eyes to adjust to the ambient lighting conditions without the amplification of the NVG. By equipping the tanker aircraft with NVIS-compatible lighting, the idea is that this adjustment time would be eliminated. Without having to wait for the receiver pilot's eyes to adjust, less time would be required for both the tanker and the receiver to complete the refueling. Thus, fewer tankers and receivers would be required to complete the same mission, or the same number of tankers and receivers would have greater capability in terms of range, endurance, or (in the case of tankers) fuel for other receivers.

To judge the magnitude of this time savings, we refer to tests conducted by the Air Force to evaluate the feasibility of AR with NVIS use (ACC, 1998). In these test refueling operations, NVG "did not notably speed up the air refueling process." The average time for receiver pilots' eyes to adjust to the ambient lighting with conventional tanker lights dimmed was only 21 seconds after raising their NVG. In our analysis, we allowed an adjustment period of one minute, giving the benefit of the doubt to the usefulness of the time savings from using NVIS-compatible lighting. This one minute of effectiveness translates to an improvement in effectiveness of 0.36 percent (see Figure 8.1).

Discounted Hypotheses

In this section, we discuss the hypothesized benefits of lights-out tanker operations at airfields, increased safety, and benefit to special operations aircraft.

Lighting upgrades to the cockpit and cargo-compartment sections of the aircraft would allow landing, cargo loading and unloading, and takeoffs from airfields operating under covert lighting conditions. For the KC-10, these types of operations would be of limited value because of the minimum runway length and materiel handling equipment necessary for KC-10 cargo operations. The performance characteristics of the KC-10 have led Air Force policy to limit the KC-10 to operating at airfields with runways that are at least 7,000 ft long and 147 ft wide. Most airports that have runways this size are also large enough to have regular civil air traffic and are located in or near urban areas with significant populations and cultural lighting. The

---

[3] The restriction on conducting AR is found in the respective aircraft operation regulations of the A-10, B-52, B-1, F-15, F-15E, F-16, F-22, and F-117. Some of the regulations allow the use of NVG to facilitate rendezvous but prohibit their use in the precontact position (generally 50 ft aft of the tanker) until refueling operations have concluded.

Figure 8.1
Change in Tanker Effectiveness for Employment with Faster Rendezvous Times

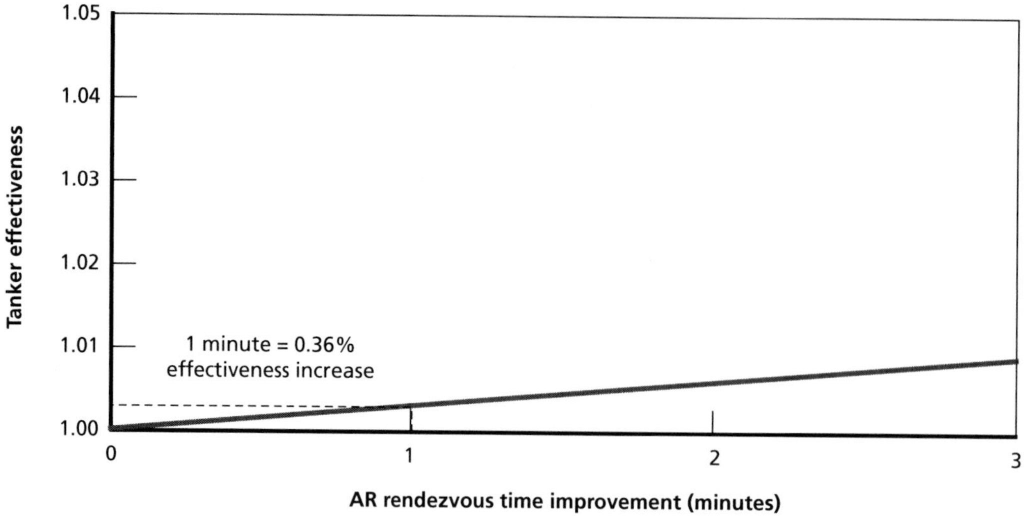

SOURCE: ACC, 1998.

RAND TR901-8.1

size and locations of these airports make it unlikely that an aircraft as large as the KC-10 would be operated unnoticed. In addition to the constraints on runway size, the KC-10 requires specialized handling equipment to load and unload cargo from its door, which is approximately 15 ft off the ground. While this equipment is not uncommon to Air Force mobility bases and large civil aviation cargo operations, it is not likely to be present at more austere locations where blacked-out operations would be feasible and desired. The operators of this ground equipment would also need to wear NVG while in very close proximity to the aircraft. The additional complexity of the operation, the potential for aircraft damage, and the low probability that operating under covert lighting conditions would effectively conceal KC-10 ground operations all limit the value of this modification option.

At the beginning of this study, we hypothesized that there was the potential to increase safety when operating in mixed-lighting environments by making all aircraft more visible to each other. To investigate this benefit, we looked at Air Force safety reports from OIF. Specifically, we analyzed Hazardous Air Traffic Report Summaries from Air Forces Central for the years 2007–2009. We also had email contact with the Air Force Forces chief of safety in country (Conaway, 2009). Both the official reporting and the discussion with the chief of safety indicated that operations in mixed-lighting environments were not as problematic as initially believed. Rather, most occurrences of near-midair collisions involved unmanned aircraft. While tanker aircraft do not currently operate at fields using these mixed-lighting conditions, reports suggests that even at fields where aircraft operate both with and without NVG, there has not been a problem. A second common factor in near misses was the unfamiliarity of aircrews with theater procedures, especially when these aircrews first rotate into theater.

Currently, the KC-10 does not routinely conduct AR missions with special operations receiver aircraft, primarily because of the performance differences between the aircraft. The KC-10's ideal speed range for refueling is above that of the special operations C-130 air-refuelable aircraft. KC-10s have, in emergency circumstances, refueled special operations aircraft, but this is out of the ordinary. Instead, specially trained, tasked, and equipped KC-135 crews and aircraft are assigned as tankers for these special operations missions. The addition of NVIS-compatible lighting to the KC-10 would not overcome the advantages of employing the KC-135 for special operations refueling missions. As a result, we do not find a benefit to the special operations mission from equipping the KC-10 with NVIS-compatible lighting.

Night Vision Imaging System–Compatible Lighting Cost-Benefit Summary

In this section, we summarize the costs and calculated benefits for upgrading the KC-10 with NVIS-compatible lighting. All costs and benefits are in terms of NPV. NPV calculations take into account the time value of money and should be thought of as incorporating associated costs and benefits over the remaining life of the aircraft.[4] Table 8.5 presents the costs of modifying the KC-10 with NVIS-compatible lighting.

NVIS-compatible lighting would provide only a marginal wartime effectiveness improvement in the employment mission. Based on the number of receivers that perform both AR and NVG operations, combined with data from the first day of OIF, we assume that 10 percent of receivers could benefit for an overall effectiveness improvement of 0.036 percent, and only in the employment mission. Table 8.6 presents the cost-benefit summary for NVIS-compatible lighting on the KC-10.

Table 8.5
Costs of NVIS-Compatible Lighting

System or Category	AUPC (FY 2009 $ millions)
Exterior	1.8
ARO station	0.8
Cockpit	1.0
Cargo area	0.03
Total	3.6

Table 8.6
NVIS-Compatible Lighting Cost-Benefit Summary

AUPC (FY 2009 $ millions)	Mission Mix	NPV Benefit (FY 2009 $ millions)
3.6	Employment-heavy	0.1
3.6	Deployment-heavy	0.0
3.6	Average benefit	0.05

---

[4]   NPV is the appropriate way to judge modernization investments based on the true resource cost. OMB directs this kind of discounted analysis. We used the current (December 2009) OMB-directed real long-term discount rate of 2.7 percent in our analysis, which represents the return on investment (OMB, 2009).

CHAPTER NINE

Conclusions

After examining the costs and benefits of each of the modernization options, we now compare the relative merits of each. We take the approach of ranking the modernization options by their cost-effectiveness ratio, as described in Chapter Four. By using the cost-effectiveness ratio, we see not only how the options compare in terms of bang for the buck but also at what point the returns on modernization spending begin to decrease. It is worth noting that the specific costs and benefits in this study result from the extent of the specific upgrade as well as our best estimates of the concrete benefits that would result from each option. However, using our parametric methodology, the relative ranking of the modernization options can be adjusted using different levels of upgrade and different assumptions regarding effectiveness improvements to determine the benefits. This approach of comparing the benefits and costs of the modernization options does not capture those costs or benefits that are inherently not quantifiable but may be an important consideration when making the decision to upgrade the KC-10 fleet. In these cases, we review these important considerations for each of the options.

The modernization options in order of the greatest to least cost-effectiveness are adding a TDL, CNS/ATM, additional multipoint refueling, defensive systems, and NVIS-compatible lighting. The first three—TDL, CNS/ATM, and additional multipoint refueling—all have positive NPVs,[1] meaning that the overall benefit is greater than the cost to procure the upgrades. Upgrades for defensive systems are cost-effective (i.e., have a positive NPV) if either (1) KC-10s are used heavily for employment missions and can be based significantly closer to AR orbit locations or (2) KC-10s are used to offset C-17s in an airlift role. NVIS-compatible lighting is not cost-effective for the KC-10. Table 9.1 shows the cost and benefit of each of the modernization options. The benefits are based on the average of two mission mixes that represent different ways in which the KC-10 could be used in wartime: one weighted toward theater employment missions, the other weighted toward deploying fighter-sized aircraft to theater. In the following paragraphs, we present the NPV of each option using the average of the values based on the employment- and deployment-heavy wartime mission mixes.

Adding a TDL to the KC-10 has the greatest cost-effectiveness ratio of the options. The data link is a relatively inexpensive upgrade compared with the other options, at less than $1 million per aircraft. In addition to other capabilities, a TDL would provide the KC-10 with position and mission information on receiver aircraft without relying on voice communication. This information would allow the reduction of planned overlap times and facilitate faster ren-

---

[1] The overall NPV of a modernization option is the benefit less the total cost, so positive values represent more benefit than cost and negative values indicate that the option costs more than the benefit it provides.

Table 9.1
Costs and Average Benefits of Each Modernization Option

Modernization Option	FY 2009 $ Millions/TAI	
	Cost	Benefit
TDL	0.7	6.5
Additional multipoint refueling capability	4.2	11.6
Defensive systems	21.4	10.2
NVIS-compatible lighting	3.6	0.1
CNS/ATM	7.5	26.1

NOTE: All costs and benefits are presented in terms of millions of FY 2009 dollars per aircraft. We express this as FY 2009 $ millions/total aircraft inventory to indicate that these per-aircraft values were calculated using the entire KC-10 fleet size.

dezvous with receiver aircraft. The fuel and time savings from the reduced planed overlap time and faster rendezvous combine to yield an NPV of $5.8 million/TAI.

Modifying the KC-10 avionics to be compliant with upcoming worldwide equipage mandates has the next-highest cost-effectiveness ratio and an average NPV of $18.6 million/TAI. Most of the benefit ($20 million/TAI) of the CNS/ATM upgrade is the avoidance of fuel penalties by having equipment that is mandated to access the most fuel-efficient altitudes. Under a broad range of assumptions regarding savings and fuel costs, the CNS/ATM upgrade is cost-effective based on peacetime savings only. The wartime missions of deployment and air bridge would have effectiveness improvements of 3.8 and 2.4 percent, respectively. The resulting NPV for the employment- and deployment-heavy mission mixes is $16.2 million and $21.0 million, respectively. Our findings show that, even under a worst-case cost scenario, the savings resulting from KC-10 fleet modernization would exceed the cost of the upgrade long before the fleet is retired in 2045. This work is detailed in *Assessing the Cost-Effectiveness of Modernizing the KC-10 to Meet Global Air Traffic Management Mandates* (Rosello et al., 2009).

Additional multipoint refueling capability also has a positive NPV, averaging $4 million/TAI, based on limiting the number of receivers in formation with the tanker to six aircraft. The primary benefit of multipoint refueling is in the employment mission when refueling multiple strike and air defense aircraft. According to KC-10 operators, the current wing refueling pods on the KC-10 are unreliable and have potential failure modes that make their use unattractive. Currently, this capability is not employed by the KC-10 in the conflicts in Afghanistan and Iraq for those reasons. If aircraft refueling formations were expanded to eight receivers per tanker, the NPV of additional multipoint refueling would increase to $7.3 million/TAI.

Defensive system upgrades are cost-effective only if these systems allow the KC-10 to be based significantly closer to operational AR locations in a conflict than established in planning documents and practiced in recent conflicts. Defensive systems may also be cost-effective if they allow the KC-10 to be used more in an airlift role, thus freeing a number of large defensive system–equipped airlifters (C-17s or C-5s, for example) to conduct other missions for which they are best suited. In the case of defensive systems and allowing closer basing, trade-offs can be made in terms of the cost and the extent of the upgrade, as well as how close the military is willing to base the aircraft. Decreased distance between the tanker base and refueling locations makes tanker aircraft more effective but may base the tankers at locations under greater threat. In terms of cost-effectiveness, it is easier to overcome the cost of a less expensive system with less capability than to overcome the costs of a more expensive and more capable set of upgrades. We present, in our estimation, a reasonable set of components for AR and airlift missions conducted by a subsonic, nonstealthy aircraft like the KC-10. Based on the defensive systems we suggest and basing the KC-10s 200 nm closer to AR orbits, we estimated an NPV of –$11.2 million/TAI. In other words, the cost of the upgrade would be more than the value of its benefit.

Retrofitting the KC-10 with NVIS-compatible lighting is not cost-effective. The NVIS upgrade is not cost-effective because there is minimal change to tanker effectiveness with the upgrade. Air Force testing and empirical safety data support the lack of improvement in tanker mission effectiveness with NVIS-compatible lighting. The NPV of the NVIS upgrade for the KC-10 is –$3.5 million/TAI

Figure 9.1 shows each of the modernization options in order of their cost-effectiveness in a cumulative plot of costs and benefits. As the figure shows, TDL, CNS/ATM, and multipoint refueling capability each provide more benefit than cost, but defensive systems and NVIS-compatible lighting cost more than the benefit they provide. As a package, if all the upgrades were pursued, the overall benefit would be greater than the overall cost for all the upgrades.

Figure 9.1
Cumulative Cost-Benefit Curve of Modernization Options

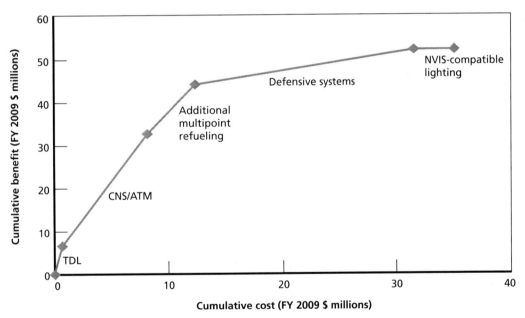

Bibliography

305th Operation Support Squadron, local KC-10 training mission operations data, spreadsheet, October–December 2007.

ACC—*see* Air Combat Command.

AFOTEC—*see* U.S. Air Force Operational Test and Evaluation Center.

Air Combat Command, *Night Air-to-Air Refueling with Night Vision Goggles Tactics and Development and Evaluation (TD&E) Final Report*, Air Combat Command Project 98-577RA, March 1998.

Air Force Doctrine Document 2-6, Air Mobility Operations, March 1, 2006.

Air Force Instruction 11-221, Air Refueling Management (KC-10 and KC-135), November 1, 1995.

————— 11-2A/OA-10, Vol. 3, A/OA-10 Operations Procedures, February 2002.

————— 11-2B-1, Vol. 3, B-1 Operations Procedures, July 20, 2005.

————— 11-2B-52, Vol. 3, B-52 Operations Procedures, June 22, 2005.

————— 11-2F-15, Vol. 3, F-15 Operations Procedures, July 21, 2004.

————— 11-2F-15E, Vol. 3, F-15E Operations Procedures, August 11, 2009.

————— 11-2F-16, Vol. 3, F-16 Operations Procedures, September 30, 2005.

————— 11-2F-22A, Vol. 3, F-22A Operations Procedures, December 8, 2009.

————— 11-2F-117, Vol. 3, F-117 Operations Procedures, November 23, 2004.

————— 11-2KC-10, Vol. 3, KC-10 Operations Procedures, January 18, 2006.

————— 65-503, U.S. Air Force Cost and Planning Factors, February 4, 1994.

Air Force Materiel Command, Office of Aerospace Studies, *Analysis of Alternatives (AoA) Handbook: A Practical Guide to Analyses of Alternatives*, Kirtland AFB, N.M., July 2008. As of July 1, 2009: http://www.oas.kirtland.af.mil/AoAHandbook/AoA%20Handbook%20Final.pdf

Air Mobility Command, Global Decision Support System Historical Mission Reporting Tool, accessed July 2008.

"Aircraft Stricken in 2003," *Naval Aviation News*, Vol. 86, No. 5, July–August 2004.

Alshtein, Alex, Theodore Cochrane, Kelly Connolly, James DeArmon, Kyle Jaranson, Mathew McNeely, Paul Otswald, Timothy Stewart, and Michael Tran, MITRE Corporation, "Simulating Civilian and U.S. Military Use of European Airspace," paper presented at the 25th Digital Avionics Systems Conference, IEEE and American Institute of Aeronautics and Astronautics, Portland, Ore., October 15–19, 2006.

ARINC Engineering Services, *Final Report for the KC-10 Aircraft Modernization Program (AMP) Concept Refinement Study (CRS)*, Annapolis, Md., March 26, 2007.

Binger, William, *Night Air-to-Air Refueling (NAAR) with Night Vision Goggles (NVG) Tactics Development and Evaluation (TD&E) Final Report*, Tucson, Ariz.: Air National Guard Air Force Reserve Test Center, Air Combat Command Project 98-577RA, December 1998.

Boeing Company, *Multipoint Aerial Refueling Boom Feasibility Study*, Technical Report ASD/XR-71-30, Wichita, Kan., January 1972.

Bolkcom, Christopher, *Air Force Aerial Refueling Methods: Flying Boom Versus Hose-and-Drogue*, Washington, D.C.: Congressional Research Service, RL32910, June 5, 2006.

"C-130 Makes Emergency Landing in Baghdad Field," *Air Force Print News*, June 27, 2008.

Camana, Peter, ViaSat, "Current and Future Military Data Links," briefing presented at Technology Training Corporation seminar, Alexandria, Va., February 23–24, 2008.

Conaway, Vernon, chief of Air Force Forces Safety, email to authors, March 27, 2009.

DHS—*see* U.S. Department of Homeland Security.

DoD and JCS—*see* U.S. Department of Defense and U.S. Joint Chiefs of Staff.

F-16.net, "US F-16 Destroyed in Ground Fire," *F-16 Mishap News*, November 12, 2008.

———, "F-16 Aircraft Database," web page, accessed August 25, 2010. As of August 25, 2010:
http://www.f-16.net/aircraft-database/F-16/

Fedarko, Kevin, and Mark Thompson, with Edward Barnes, Ann Blackman, Greg Burke, Dan Cray, and Douglas Waller, "Rescuing Scott O'Grady: All for One," *Time Magazine*, June 19, 1995.

Federal Aviation Administration, *Aeronautical Information Manual: Official Guide to Basic Flight Information and ATC Procedures*, July 31, 2008.

Federal Aviation Administration Order 7610.4K, Special Military Operations, February 19, 2004.

Girard, Mary M., and Stephen W. Hill, *Dual Use of Commercial Avionics Data Links for the U.S. Air Force*, Bedford, Mass.: MITRE Corporation, October 1999.

Gonzales, Daniel, John Hollywood, Gina Kingston, and David Signori, *Network-Centric Operations Case Study: Air-to-Air Combat With and Without Link 16*, Santa Monica, Calif.: RAND Corporation, MG-268-OSD, 2005. As of September 24, 2010:
http://www.rand.org/pubs/monographs/MG268/

Greer, W. L., H. S. Balaban, W. C. Devers, G. M. Koretsky, and H. J. Manetti, *C-130 Avionics Modernization Program Analysis of Alternatives (C-130 AMP AOA)*, Alexandria, Va.: Institute for Defense Analyses, P-3589, March 2001.

Griffis, Stanley E., and Joseph D. Martin, *Development and Analysis of a Dual-Role Fighter Deployment Footprint Logistics Planning Equation*, Wright-Patterson AFB, Ohio: Air Force Institute of Technology, September 1996.

Hunsuck, John Davis, Jr., *Comparing the Effectiveness of Two KC-10 Concepts of Operation: An Examination of Tanker/Airlift Support in a Fighter Deployment to Europe*, Wright-Patterson AFB, Ohio: Air Force Institute of Technology, June 1986.

Hura, Myron, Gary McLeod, Eric V. Larson, James Schneider, Daniel Gonzales, Daniel M. Norton, Jody Jacobs, Kevin M. O'Connell, William Little, Richard Mesic, and Lewis Jamison, *Interoperability: A Continuing Challenge in Coalition Air Operations*, Santa Monica, Calif.: RAND Corporation, MR-1235-AF, 2000. As of September 24, 2010:
http://www.rand.org/pubs/monograph_reports/MR1235/

JCS—*see* U.S. Joint Chiefs of Staff.

Joint Air Power Competence Centre, *Future of Air-to-Air Refuelling in NATO*, Kalkar, Germany, June 11, 2007.

Kalt, Dex H., briefing at the Aerial Refueling Systems Advisory Group International Conference, Jacksonville, Fla., May 6, 2004.

Keaney, Thomas A., and Eliot A. Cohen, *Gulf War Air Power Survey Summary Report*, Washington, D.C.: U.S. Government Printing Office, 1993.

Kennedy, Michael, Laura H. Baldwin, Michael Boito, Katherine M. Calef, James S. Chow, Joan Cornuet, Mel Eisman, Chris Fitzmartin, Jean R. Gebman, Elham Ghashghai, Jeff Hagen, Thomas Hamilton, Gregory G. Hildebrandt, Yool Kim, Robert S. Leonard, Rosalind Lewis, Elvira N. Loredo, Daniel M. Norton, David T. Orletsky, Harold Scott Perdue, Raymond A. Pyles, Timothy Ramey, Charles Robert Roll, Jr., William Stanley, John Stillion, Fred Timson, and John Tonkinson, *Analysis of Alternatives (AoA) for KC-135 Recapitalization: Executive Summary*, Santa Monica, Calif.: RAND Corporation, MG-495-AF, 2006. As of September 24, 2010:
http://www.rand.org/pubs/monographs/MG495/

Killingsworth, Paul, *Multipoint Aerial Refueling: A Review and Assessment*, Santa Monica, Calif.: RAND Corporation, DB-152-CRMAF, 1996. As of September 24, 2010:
http://www.rand.org/pubs/documented_briefings/DB152/

Knisely, Mike, "KC-10 CNS/ATM Review," briefing to KC-10 Tiger Team Meeting, Wright-Patterson AFB, Ohio, January 31, 2008.

L-3 Communications Integrated Systems, *KC-10 Aircraft Modernization Program (AMP) Concept Refinement Study, Scientific and Technical Report*, March 15, 2007, not available to the general public.

MacGregor, Timothy B., *Reassessing USAF Tanker Employment CONOPS and Command and Control to Maximize Air Mobility Capacity*, Newport, R.I.: Naval War College, February 9, 2004.

NATO—*see* North Atlantic Treaty Organization.

"Navy Fighter Jets Crash in Persian Gulf," Associated Press, January 7, 2008.

"Navy's F-14A Tomcat Crashes in Iraq," Associated Press, April 2, 2003.

North Atlantic Treaty Organization, "Air-to-Air Refuelling," ATP-56(B), January 22, 2010.

Northrop Grumman, "Zone of Vulnerability," web page, undated. As of September 24, 2010:
http://www.es.northropgrumman.com/countermanpads/threat/zone.html

Office of the Assistant Secretary of Defense (Public Affairs), "Contracts," press release, No. 0075-08, January 29, 2008. As of September 24, 2010:
http://www.defense.gov/contracts/contract.aspx?contractid=3696

Office of Management and Budget, "Discount Rates for Cost-Effectiveness, Lease Purchase, and Related Analyses," Appendix C in *Guidelines and Discount Rates for Benefit-Cost Analysis of Federal Programs*, Circular A-94, revised December 2009. As of September 24, 2010:
http://www.whitehouse.gov/omb/circulars/a094/a94_appx-c.html

Office of the Under Secretary of Defense (Comptroller), *National Defense Budget Estimates for FY 2009* [Green Book], Washington, D.C., March 2008. As of September 24, 2010:
http://comptroller.defense.gov/budget2009.html

———, *FY 2011 Amended President's Budget: Aircraft Procurement Modifications, Air Force*, 2010. As of November 16, 2010:
http://www.saffm.hq.af.mil/shared/media/document/AFD-100128-071.pdf

OMB—*see* Office of Management and Budget.

Payne, Kevin M., *Broadband to the Battlefield*, Maxwell AFB, Ala.: Air University, Air Command and Staff College, April 2006.

Pinter, Maj Bill "Pinto," Air Combat Command/XPS (Strategy, Concepts, and Doctrine), "CAF 2025 Flight Plan," briefing and accompanying notes, February 2004.

Pruitt, James, Air Mobility Command, "Tankers and Transloading: Making the Most of Dual Role Tankers," briefing presented at the 17th Military Operations Research Society Symposium, Annapolis, Md., June 12–14, 2007.

Rockwell Collins, Boeing Company, and Honeywell International, *KC-10 Aircraft Modernization Program (AMP) Concept Refinement Study: Final Report*, March 2007, not available to the general public.

Rosello, Anthony D., Sean Bednarz, Michael Kennedy, Chuck Stelzner, Fred Timson, and David T. Orletsky, *Assessing the Cost-Effectiveness of Modernizing the KC-10 to Meet Global Air Traffic Management Mandates*, Santa Monica, Calif.: RAND Corporation, MG-901-AF, 2009. As of September 24, 2010: http://www.rand.org/pubs/monographs/MG901/

Schroeder, Matt, "Rogue Missiles: Tracking MANPADS Proliferation Trends," *Jane's Intelligence Review*, November 1, 2007.

"Serb Discusses 1999 Downing of Stealth," Associated Press, October 26, 2005.

Smiths Aerospace, *KC-10 Aircraft Modernization Program Concept Refinement Studies: Cost Analysis Guidance and Procedures, Final Report*, March 12, 2007, not available to the general public.

Stillion, John, David T. Orletsky, and Chris Fitzmartin, *Analysis of Alternatives (AoA) for KC-135 Recapitalization: Appendix B—Effectiveness Analysis*, Santa Monica, Calif.: RAND Corporation, 2006, not available to the general public.

UK Ministry of Defence, *Military Aircraft Accident Summary: Aircraft Accident to Royal Air Force Hercules ZH876*, undated.

———, *Military Aircraft Accident Summary: Aircraft Accident to Royal Air Force Tornado GR MK4A ZG710*, May 2004.

———, "RAF 'Determined to Learn the Lessons' of Hercules XV179," *Defence Policy and Business*, October 22, 2008.

USAF—*see* U.S. Air Force.

U.S. Air Force, *TO 1C-17A-1-1 Flight Manual Performance Data, USAF Series C-17A*, January 1, 2005a.

———, *TO 1C-10(K)A-1-1 Flight Manual Performance Data, USAF Series KC-10A*, February 1, 2005b.

———, *Initial Capabilities Document for Air Refueling*, April 27, 2005c, not available to the general public.

———, *Executive Summary: Aircraft Accident Investigation, F-16 CG, S/N 90-0776, 524th Expeditionary Fighter Squadron (EFS)*, Balad Air Base, Iraq, November 27, 2006.

———, *TO 1C-10(K)A-1 Flight Manual*, Vol. 1, Series KC-10A, June 1, 2007.

U.S. Air Force, Electronic Systems Center, Strategic Projections of Airspace Requirements and Certifications (SPARC), software tool and database, undated. As of January 7, 2009: https://sparc.qdmetrics.com/SPARC/home.jsp

U.S. Air Forces Central, *AOR HATR Updates*, October 15, 2008, through November 25, 2009.

U.S. Air Force Operational Test and Evaluation Center, "LAIRCM: Protecting the Military's Big Birds," August 1, 2007. As of November 16, 2010: http://www.afotec.af.mil/news/story.asp?id=123057893

"US C-5 Cargo Plane Hit by Missile on Takeoff from Baghdad," Agence France-Presse, January 8, 2004.

U.S. Central Command, "'Aircraft Mishap' in Northern Iraq—CENTCOM," Release No. 04-12-64, December 30, 2004.

U.S. Department of Defense, "DoD Identifies Air Force Casualty," Release No. 754-07, June 16, 2007.

U.S. Department of Defense and U.S. Joint Chiefs of Staff, *Mobility Capabilities Study*, Washington, D.C., 2005, not available to the general public.

———, *Mobility Capabilities and Requirements Study 2016*, Washington, D.C., February 26, 2010, not available to the general public.

U.S. Department of Homeland Security, Science and Technology Directorate, *Counter-MANPADS Program Summary*, Washington, D.C., July 2006.

"U.S. F-16 Warplane Crashes in Iraq, Pilot Uninjured," Xinhua News Agency, July 16, 2007.

U.S. Joint Chiefs of Staff, *Air Mobility Operations*, Washington, D.C., Joint Publication 3-17, October 2, 2009.

U.S. Navy Public Affairs, "Second Pilot Identified in F/A-18 Crash," press release, May 5, 2005.

Wilson, Clay, *Network Centric Operations: Background and Oversight Issues for Congress*, Washington, D.C.: Congressional Research Service, RL32411, March 15, 2007.